CIRCULATING COPY
Sea Grant Depository

LOAN COPY ONLY

The Hazards of Diving in Polluted Waters

Proceedings of an International Symposium
held in Bethesda, Maryland
Dec. 13-16, 1988

Symposium Sponsored by

Maryland Sea Grant College
NOAA National Undersea Research Program
Undersea and Hyperbaric Medical Society

A Maryland Sea Grant Publication
College Park, Maryland

Published by the Maryland Sea Grant College under grant numbers NA 86AA-D-SG 006 and NA90 AA-D-SG 063 from the National Oceanic and Atmospheric Administration.

Publication Number
UM-SG-TS-92-02

Copyright © 1992 Maryland Sea Grant College
Printed in the United States of America

For more information on Maryland Sea Grant publications, write:

Maryland Sea Grant College
University of Maryland
College Park, Maryland 20742

Printed on recycled paper

The following papers in this volume first appeared in *Undersea Biomedical Research: The Journal of the Undersea and Hyperbaric Medical Society*, Bethesda, Maryland. They are reprinted here with permission:

R. R. Colwell, Microbiological hazards: background and current perspectives; G. Losonsky, Infections associated with swimming and diving; M. T. Kelly, Pathogenic *Vibrionaceae* in patients and the environment; B. F. Masterson, Protection of recreational divers against water-borne microbiological hazards; W. Schane, Prevention of skin problems in saturation diving; D. M. Rollins, Diving equipment as protection against microbial hazards; J. E. Amson, Protection of divers in waters that are contaminated with chemicals or pathogens; R. Thompson, Diving in nuclear power plants; J. M. Wells, Hyperthermia in divers and diver support personnel; Y. S. Park and S. K. Hung, Physiology of cold-water diving as exemplified by Korean women divers; M. L. Nuckols, M.W. Lippitt and J. Dudinsky, The liquid-filled suit-intersuit concept: passive thermal protection for divers; W. C. Phoel and J. M. Wells, Evolution of the National Oceanic and Atmospheric Administration's capabilities for polluted water diving; S. M. Barsky, Practical systems for contaminated water diving; M. McGovern and G. Roberts, Safe and cost-effective methods of diving in contaminated water in the Midwest; J. P. Imbert, Y. Chardard and P. Poupel, Intervention in hostile environments: Comex experience. Part of S. W. Joseph's paper also appeared as Aquatic and terrestrial bacteria in diving environments: monitoring and significance.

Contents

Preface	v
The National Undersea Research Program	vii
Introductory Remarks: The Ocean as an Alien Environment	1

Section 1. Microbiological Hazards

Microbiological Hazards: Background and Current Perspectives, *Rita R. Colwell*	13
Diving Equipment as Protection against Microbial Hazards, *David M. Rollins*	19
Monitoring and Significance of Aquatic and Terrestrial Bacteria in the Aquatic Environments of Divers, *Sam W. Joseph, J.B. Conway and S.G. Kalichman*	23
Pathogenic *Vibronaceae* in Patients and the Environment, *Michael T. Kelly*	41
Protection of Recreational Divers against Water-borne Microbiological Hazards, *B.F. Masterson*	49
Prospective Study of Diving-Associated Illnesses, *Rita R. Colwell, Anwar Huq, Kelly A. Cunningham and Genevieve Losonsky*	63

Section 2. General Contamination

Evolution of the National Oceanic and Atmospheric Administration's Capabilities for Polluted Water Diving, *William C. Phoel and J. Morgan Wells*	73
Intervention in Hostile Environments: Comex Experience, *J.P. Imbert, Y. Chardard and P. Poupel*	85

Hyperthermia in Divers and Diver Support
Personnel, *J. Morgan Wells* — 93

Physiology of Cold-Water Diving as Exemplified by
Korean Women Divers, *Y.S. Park and Suk Ki Hong* — 97

The Liquid-filled Suit/Intersuit Concept: Passive
Thermal Protection for Divers, *M.L. Nuckols,
M.W. Lippitt and J. Dudinsky* — 121

Section 3. Chemical Hazards

Protection of Divers in Waters Which Are
Contaminated with Chemicals or Pathogens,
J.E. Amson — 133

Section 4. Equipment and Procedures

Practical Systems for Contaminated Water Diving,
Steven M. Barsky — 149

Diving in Nuclear Power Plants, *Randy Thompson* — 157

Safe and Cost-Effective Methods of Diving in
Contaminated Water in the Midwest, *Mike McGovern
and Greg Roberts* — 163

Section 5. Diving Medicine

Infections Associated with Swimming and Diving,
Genevieve Losonsky — 171

Prevention of Skin Problems in Saturation Diving,
William Schane — 181

Hypothermia, *Pierre D'Hemecourt* — 185

Unsolved Problems of Undersea Living,
Lawrence W. Raymond — 193

Preface

Increasingly, as divers work in and explore the marine environment, they confront an expanding array of contaminants and other threats to their health. These dangers include microbes from sewage and other sources, chemicals, cold, heat and even radioactivity.

Divers are called upon to function under a wide range of conditions, from inside nuclear power facilities to deep within pools that may be contaminated by chemicals, sewage or other toxic substances. We are learning that even environments previously thought to be benign--such as coastal ocean waters--may pose some hazards, especially near sewage outfalls or other areas which feel the effects of human habitation.

In this volume a number of researchers discuss the many hazards divers are likely to encounter in polluted waters. These papers were originally presented at an international conference convened in Bethesda, Maryland, by the National Oceanic and Atmospheric Administration, the Undersea and Hyperbaric Medical Society, and the University of Maryland Sea Grant College. In addition to descriptions of microbial and chemical hazards, the proceedings also includes introductory remarks by ocean pioneer John Craven, a section on diving equipment and procedures for decontamination, and a section on diving medicine. Many of the scientific papers presented here were originally published in the journal of the Undersea and Hyperbaric Medical Society, and Maryland Sea Grant gratefully acknowledges the help and cooperation of Dr. Hugh Van Liew, the editor of that journal, as well as Ms. Anne Barker of the Society.

The planning committee for the hazardous diving conference included Dr. Rita R. Colwell and Dr. Jack Greer, University of Maryland; Dr. William Busch, National Oceanic and

Atmospheric Administration; and Dr. Leon Greenbaum, Undersea and Hyperbaric Medical Society. Mr. Mark Jacoby, of Maryland Sea Grant, helped plan and arrange the conference. The planning committee wishes to thank Ben Haskell, Mary Moynihan and Elaine Friebele for assistance with the manuscript and everyone who helped make the conference and this proceedings a reality.

The National Undersea Research Program

The National Oceanic and Atmospheric Administration (NOAA), under the aegis of its National Undersea Research Program (NURP), directly assists a large cadre of marine research scientists to conduct their scientific activities under the sea. This research is accomplished using manned submersibles, remotely operated vehicles, and compressed-air scuba, mixed-gas, and saturation mode diving. Additionally, NURP assists all divers of the nation through research undertaken in accordance with the terms of Sec. (21)(e) of the Outer Continental Shelf Lands Act of 1978 (PL 95-372; 43 USC 1331 *et seq.*). This statute requires NOAA, under the authority delegated by the Secretary of Commerce, to "conduct studies of underwater diving techniques and equipment suitable for protection of human safety and improvement of diver performance."

The Technical Report series published by NURP is intended to provide the marine community with results of undersea research, often presented at NURP-sponsored symposia and workshops, in a timely fashion. In the majority of instances, participants at symposia or workshops are reporting on results of NURP-sponsored research. In such instances, the printing of their papers meets report requirements of grantees to the Office of Undersea Research. In other instances, the topic is of direct interest to NURP and could represent a framework for future research if published and distributed to the science community. Results reported in NURP's Technical Report series may be preliminary or require further development, refinement, or validation, and this research may be beyond the scope or mission of NURP. Reports in this series do not carry any endorsement or approbation on the part of NURP, nor can NURP accept any liability for damage resulting from incorrect or incomplete information.

The papers presented in this volume are the result of an international symposium cosponsored by NURP, the Maryland Sea Grant Program and the Undersea and Hyperbaric Medical Society and held in Bethesda, Maryland, December 13-16, 1988. The purpose of the International Hazardous Diving Symposium was to establish and describe the present level of knowledge and procedures of diving under such conditions. This joint report, published by the Maryland Sea Grant Program, therefore represents the status of diving in hazardous environments.

Papers are printed essentially as presented, and publication is in partial fulfillment of requirements under Sea Grant award NA-86-AA-D-SG006. Comments on the report are welcome. They should be directed to:

> National Undersea Research Program
> NOAA
> 1335 East-West Highway, Room 5262
> Silver Spring, MD 20910

> David B. Duane
> Director, NURP

Introductory Remarks: The Ocean as an Alien Environment

JOHN P. CRAVEN
Law of the Sea Institute
University of Hawaii
Honolulu, Hawaii

In the halcyon days of the deep submergence programs great strides were made in the development of small deep diving submersibles as well as in the ability of humans to descend, as swimmers, to great depths in the sea. Both deep diving communities were aware of the strides that the other was making in extending the range of humans in the sea. So it was, that an observer on a small submersible operating in about 600 feet of water came upon a swimmer at the same depth. The submariner, marveling at this phenomenon, maneuvered his craft so that he came face-to-face with the swimmer who peered at him through the porthole. "Hello," said the submariner. "What are you doing out there?" Through the port came the muffled answer, "I am drowning you idiot."

It has ever been the case for humans, as they seek to return to their evolutionary roots in the ocean, that they find that some aspect of the watery environment is dangerous to their health, if not lethal. This has been a great disappointment, since those of us who have been involved with "man in the sea" have been aware that we still possess the fundamental characteristics which ought to permit the 'naked human' to disport with the whales and the dolphins to the deepest part of the ocean.

Certainly, at one time our primeval predecessors could frolic in the primeval sea, but as humans evolved out of the sea and became more terrestrial, the physical characteristics of the ocean environment, such as salinity, temperature and pressure, became virtual toxic elements preventing us from rediscovering our roots. Despite our ancient ties to the ocean, it became a hostile and unfriendly place for man. Is this really a valid hypothesis? Let us test it by examining the successive introduction of those toxic elements that now limit man's extension into the sea.

Mammal studies have already demonstrated that we may flood the lungs of animals with Ringers solution (solution that has a salinity that is isotonic with blood) that is saturated with oxygen. Such mammals can breathe, even though their lungs are flooded, and as free-flooded creatures — almost totally liquid — they can oscillate from the surface to depths as great as 1000 meters without suffering the bends or air embolism. If the fluid temperature is elevated and close to body temperature, there is no thermal heat loss and the animal could survive indefinitely were it not for the build up of CO_2 in the body. Presumably (although this has not yet been accomplished) a shunt could be placed in the venous system and the blood passed through a CO_2 dialysis machine which purges CO_2 from the system.

Numerous animal studies have been conducted at Duke University in which the animals have had their lungs filled with fluid and which have been dewatered before the CO_2 has risen to unacceptable levels. A number of emphysema patients have had one lung flooded for therapeutic purposes, and at least one healthy volunteer has had a single lung flooded for experimental purposes. The indications are thus strong that an otherwise naked human fitted with an "artificial gill" could survive in a sea of Ringers solution and dissolved oxygen.

What happens if we now substitute sea water for the Ringers solution? The divergence of evolutionary development and environmental change has introduced at least two major pollutants: potassium and cold. The salinity of human blood was indeed isotonic with the salinity of the ocean many millions of years ago.

Today the salinity of the ocean is relatively high in potassium and other salts. As a consequence, many victims of drowning are not the victims of oxygen deprivation but are the victims of potassium shock as the potassium ion migrates across human membranes into the central nervous system.

The low temperature of the water and its high thermal conductivity also conducts body heat away and soon induces thermal shock. This can be mitigated by thermally heated wet suits. Thus, the current theoretical limit for humans in the sea is that of a human immersed in an oxygen-fed, oxygen-rich Ringers solution, which is separated from the sea by a pressure-compensated envelope, the human requiring a shunt in the veins leading to a dialysis machine and wearing a heated wet suit. Such a one ought, in principle at least, to be able to oscillate between the surface and depths of at least 1000 meters without concern for bends or embolism or other physiological dysfunction.

How far have we come in meeting that ideal adaptation? We have not yet a flooded lung, for we are reluctant to introduce the artificial gill. The alternative is the breathing of low density gases so that the carbon dioxide product can be exhaled and the gas removed by external scrubbers. In early 1988, the French through COMEX and the Navy conducted open sea dives demonstrating a "man in the sea" capability at depths of more than 500 meters. The divers in this experimental program breathe a mixture of hydrogen and oxygen. The layman finds this exciting, and to a first order unbelievable, for he correctly perceives that hydrogen and oxygen are hypergolically explosive mixtures at atmospheric pressure. At the pressures of the deep, however, it is not possible to sustain ignition, much less entertain explosion with such mixtures, and hydrogen is the lightest of all possible diluent gases. There must be a diluent, for the partial pressure of oxygen must be maintained at less than one atmosphere if oxygen toxicity is to be avoided.

At the outset many of us had believed that the limit of depth for divers was to be determined by the physiological work involved in the breathing of high density gases. Such may yet be the case. Our initial solution in the Sea Lab experiments was the breathing of helium and oxygen mixtures. The limits of this gas proved, however, to be neuro-physiological, with a phenomenon labeled High Pressure Nervous Syndrome wherein the diver at deep depth enters a phase of disorientation and convulsive reaction which if not mitigated can lead to death. A number of techniques for adaptation and avoidance have ben introduced, initially the use of staged compression allowing the body to adapt to the high pressure gas condition. The introduction of nitrogen into the gas mixture was also a palliative. Chamber dives using TRIMIX have been successfully conducted to depths of 600 meters with humans and to depths of 1000 meters with animals. But the transfer of breathing gas from some appropriate Nitrox mixes to hydrogen-oxygen appears to be the most effective way for mitigating this syndrome. Thus we can now keep the diver alive and well in his heated wet suit, breathing Hydrox mixtures with his breathing apparatus, his tissues saturated with nitrogen and hydrogen, his carbon dioxide exhaled into a carbon dioxide scrubber and his ambient pressure maintained in a habitat or personnel transfer capsule. The latter is a pressure-contained closed environment in which one can regulate the rate at which gas mixtures and pressures are changed, so that the body can adapt to each new depth environment.

As with any change in environment, there were and are many other long-term and short-term effects. Bone necrosis, loss of hearing, permanent damage of vital organs by oxygen toxicity (a literal oxidation of body tissue) and susceptibility to bends are among the most prominent, but surprisingly, at least initially, many positive effects were also realized. The perfusion of tissues of the extremities with oxygen and the supply of oxygen to sclerotic tissues in the brain have had dramatic therapeutic effect. Although in the United States the use of hyperbaric medicine is marginal at best and is practiced by only a few institutions, it has become a major therapy in Japan, employed successfully for the successful treatment of gas gangrene, osteomyelitis, arteriosclerosis, diabetic ulcers, carbon monoxide poisoning, brain injury and sudden hearing loss. In addition, substantial anecdotal evidence exists for the value of hyperbaric oxygen in reversing or retarding various aging and senility processes (not Alzheimer's), and such notables as Barry Goldwater and Michael Jackson are devotees of this treatment. Indeed, your speaker fully intends to avail himself of this therapy as soon as he detects a negative partial derivative in his mental processes.

There are other environmental factors which are deemed to be hostile in the ocean environment that are in fact beneficent.

Salinity makes the ocean highly conductive, and as a result cathodic protection is required on submarines and other devices permanently immersed in the water. Some years ago the Academy of Engineering suspected that underwater electric arc welding might be a hazardous occupation. The average lay person is all too familiar with the lethality of the electrical appliance in the bathtub. But as an elementary analysis of the physics of the problem indicates, it is almost impossible to provide a lethal shock to an individual who is fully immersed in a saline medium. Navy attempts to develop electric shields against underwater saboteurs have all failed. The salinity of the water transmits electrons in such a manner as to minimize voltage gradients. The Academy could find no instance of electrocution for an immersed diver. It was found that heavy currents did indeed surge through the body and that when divers opened their mouths and separated their teeth sparks would jump across the gap. The electrolytic action was such that many underwater welders have lost the fillings in their teeth but few have suffered any other physiological effect.

Other physical characteristics of the ocean medium are in fact beneficial to the extent that there are classes of people, most of whom are deemed to be handicapped, who perform better in an aquatic environment than on land, sometimes better than physically "normal" people. Most obvious is the superior performance of blind or sight impaired divers operating in environments of limited or no visibility.

A second example is that of a pair of deaf divers operating as a team in waters of good visibility. The superiority of performance of such a team needs no demonstration. Less obvious is the performance of individuals who are double amputees of the lower limbs. The hydrodynamic efficiency of such individuals as swimmers has been demonstrated in laboratory tests. Less efficient but relatively effective as compared with their land performance are paraplegics and single and multiple amputees. Indeed all that is required for successful propulsion as a swimmer is a single operating "flipper."

Other individuals who benefit from the water environment include those with spina bifida, cerebral palsy due to the fluid damping of athetoid movements. Individuals requiring a daily dose of hyperbaric oxygen as compensation for arteriosclerosis do not need the services of expensive compression chambers to obtain this therapy. Finally, and of particular interest to this group, are those individuals who are incontinent. The discharge of human waste into the ocean with appropriate care should have no significant pollution effects on the environment. More about this later on in the paper.

The bouying effect of the salt water makes the performance of these "special" individuals even more remarkable. One cannot fall in the water. One can change altitude with ease, leaping high walls with no effort. Individuals employing buoyancy lift devices can manhandle very large heavy components, such as well casings, valves, pipes, etc. The Sea Lab experiments with which I was involved so many years ago demonstrated the ease with which individuals could construct major underwater structures without benefit of major construction tools.

What does this suggest? To some of us it suggests economically viable coastal communities in the tropics whose workers are involved with the harvest of precious black coral, the undersea mariculture of pearls, oysters, reef fish, seaweeds and other marine farm products, the maintenance and furnishing of services for underwater parks and habitats. Such a community would provide a critical mass of workers for the effective location of the clinical services required by these "special people," regardless of their location, and of course, the entry and exit devices which carry these individuals into the water for their daily avocation and recreation.

Numerous attempts have been made to initiate a Sea Grant project to explore this possibility. In each instance an enthusiastic individual has begun the start up studies only to have them frustrated by apathy, disinterest, and outright rejection, by the community who would be the primary beneficiaries of such a project. While no specific reason can be identified for this rejection and while all of us who labor in the oceanic world are accustomed to the adoption of the ocean solution as the last resort of a society that can think of no viable, even inferior, alternative, a major factor is a national phobia and obsession with oceanic hazard and oceanic pollution. Even oceanic specialists can become traumatized by the uncertainty of physiological hazards in this alien environment. It was a source of educational surprise to me to encounter this phenomenon during the SeaLab II experiments. Astronaut/Aquanaut Scott Carpenter was stung by a scorpion fish while in a saturated condition at 300 feet. Medical doctors were concerned that the atmospheric treatment in the form of histamines and antibiotics would be contraindicated at depth and there were those of the medical profession who advocated his rapid return to atmospheric pressure before commencing treatment. Others felt that he could be treated at depth. The medical doctors could not agree, and so the matter was referred to me by virtue of my command responsibility and not by virtue of my expertise. My command decision was to keep him at depth, give him a shot of antihistamines, and tell him not to think about it.

But this conference and this community has been called upon to think about it. The suggestion previously made that incontinent individuals can operate in the ocean environment without the necessity for wearing diapers has been offensive to most lay persons who have been exposed to the suggestion. It may be a concern to some professionals in this audience. Indeed society and your speaker have been struggling with the problem of sewage disposal and waste disposal in the ocean ever since Rachel Carson, nay ever since Ibsen and his play *An Enemy of the People*.

Rita Colwell was kind in not introducing me by my well-known appellation in Hawaii as the Dean of Marine Pollution. Serious studies at the University of Hawaii have indeed indicated that, properly treated and distributed, nutrients from human wastes can beneficiate the ocean environment. Other studies have shown that from an epidemiological standpoint pollution of fresh water is a real hazard to swimmers, whereas the epidemiology has been, until recently, absent for the ocean environment. The many infections which are experienced on the coral reef are indeed primarily infections resulting from coral cuts and abrasions and the deposition of coral planulae in these wounds. Sea Grant studies at the University of Hawaii have indicated that survival of viruses and biological pathogens in tropical water is of short duration.

My dermatologist recommends swimming in the ocean with open cuts and sores as an antiseptic measure. But it is my new understanding that current evidence available at this conference is that this happy situation is at best a dangerous oversimplification. Certainly fish viruses and pathogens do survive in the ocean, and there is nothing so unique about these organisms to suggest that nature can differentiate between these animal pathogens and human pathogens, and it is an ignorant individual indeed who is not aware of the propensity for evolution and adaptation of bacteria and viruses in time scales that are short in comparison with human intervention measures. Rabbits in Australia are living proof, as are the many victims of AIDS and other rapidly mutating forms of the HIV viruses.

What then is my message? In a seashell it is simply this: almost all of the physical and biological substances in the ocean, including the water itself, are alien to *Homo sapiens* as he now is, but most of the physical and biological substances in the ocean were once beneficial to that evolutionary creature from which we evolved. With care and with the development of the appropriate prosthesis and prophylactics we can protect and readapt the human to blend with the ocean environment in a manner that its enjoyment resolves many of the problems of modern man. Look therefore not only at the hazards of substances, all substances, in the ocean, but look for their potential beneficial effect--the identification of each is vital for the wise use of this great world resource.

One of these special people to whom I have alluded has already demonstrated the potential of the ocean for such individuals and for us all. One of the English speaking world's greatest poets, George Gordon, better known as Lord Byron, was born with a congenitally deformed foot. Although he tried to compensate for his lameness by engaging in boxing and cricket, he found full release for his athletic propensities in oceanic swimming. He knew the ocean as few in his century or this know the ocean. In 1812 he produced these lines, which speak eternally for all of us who also know and love the ocean:

> *And I have loved thee Ocean! and my joy*
> *Of youthful sports was on thy breast to be*
> *Borne, like thy bubbles, onward: from a boy*
> *I wantoned with thy breakers, -- they to me*
> *Were a delight; and if the freshening sea*
> *Made them a terror, 'twas a pleasing fear;*
> *For I was as it were a child of thee,*
> *And trusted to thy billows far and near,*
> *And laid my hand upon thy mane, -- as I do here.*

Section 1

Microbiological Hazards

Microbiological Hazards: Background and Current Perspectives

RITA R. COLWELL
Maryland Biotechnology Institute
University of Maryland
College Park, Maryland

Until 1980, there was a great deal of skepticism about whether or not there were health problems associated with diving in polluted waters. From 1970 to the present, however, anecdotal information began to accumulate: cases of divers suffering unusual rashes after exposure to water at sewer outfalls, a New York Harbor diver contracting meningitis, and a variety of infections purported to be associated with exposure to water at polluted dive sites. In late 1970, a workshop discussed the accumulating anecdotal evidence of microbiological hazards to divers. The number of people interested in the subject has grown significantly, and evidence is rapidly accumulating that there are, indeed, microbiological hazards to divers in polluted waters.

Although an isolated case may not appear to be significant, parameters and general principles associated with such incidents need to be extracted to understand environmental conditions and pathogens causing the infections for risk analysis and to develop preventive measures for the diving community. Even if the risks should prove to be small, they should be measured and analyzed. At present, there are no standards, particularly for recreational diving, and microbiological standards set for shellfish harvesting are, generally speaking, those employed for recreational swimming.

Specific hazards facing divers need definition. The kinds of materials discharged into waters where divers enter also need characterization. The kinds of organisms present in polluted waters should be identified as well. A useful example is a study of an ocean dump site off the coast of Puerto Rico carried out several years ago (1,2).

A variety of microbiological parameters were determined for surface waters in and around the Puerto Rico dumpsite, which was used for disposal of pharmaceutical wastes. Specific activities of microbial populations were derived from comparisons of activity measurements (uptake of radiolabeled substrates or substrate-responsive cell numbers determined by epifluorescent microscopy) and total cell numbers. Highest values were observed in samples from stations in or near the dumpsite. Similarly, largest numbers of colony-forming bacteria, enumerated on marine agar, were obtained in the vicinity of the dumpsite.

Total colony-forming bacteria were enumerated at all stations, using several different culture media, and randomly selected isolates were identified to develop diversity indices for the culturable bacterial community. Bacteria isolated on marine agar were found to be predominantly members of the *Vibrio/Aeromonas* group, with the more typical marine pseudomonads comprising less than 9% of the community. In the vicinity of the dumpsite, large numbers of Gram-positive bacteria, that is, micrococci, staphylococci, and bacilli, were recovered from water samples plated on marine agar, as well as from those plated on plate count agar, which selects against bacteria requiring sea salts for growth.

Results obtained, in particular diversity index measurements and the persistence of culturable, waste-specific organisms at the dumpsite, suggested that alterations in the natural microbial populations of surface waters of the Puerto Rico dumpsite and environs had occurred. There appeared to be a decrease in naturally occurring marine microorganisms and an increase in non-marine microorganisms.

In another study, the frequency of occurrence of plasmids in bacteria in seawater was studied (3). Plasmids conferring antibiotic resistance in bacteria found in sewage effluent at an outfall off Ocean City, Maryland, were present at significantly higher incidence than in bacteria at sites free of sewage effluent. Bacteria commonly isolated from seawater carried few plasmids and demonstrated a relatively low level of antibiotic resistance, compared to bacteria from the sewer outfall and sewage-impacted seawater offshore.

Thus, several kinds of effects at the microbiological level, can be measured. One is a change in microbial community structure and another is an increase in potential pathogens persisting in the viable but nonculturable state in seawater (4). An increased incidence of antibiotic resistance associated with potential pathogens in polluted waters may also occur, notably when an elevated concentration of heavy metals and toxic chemicals is present in the polluted water.

In a separate study carried out in Puerto Rico (2), a sewage plant discharging offshore near a bathing beach showed larger numbers of bacteria than normally present in seawater. Microbiological effects of non-disinfected, combined chemical and domestic wastewater effluent discharge into coastal waters north of Barceloneta, Puerto Rico, were investigated by membrane-filter enumeration of fecal indicator bacteria and by enrichment isolation of specific pathogens. A wastewater plume was detected and delineated around the sewage outfall, located 800 m offshore and 30 m below the ocean surface. Fecal coliform and fecal streptococcus counts within the plume were significantly larger than comparable counts outside the plume. The shape and location of the plume was compatible with current patterns and prevailing winds and extended as far as 3.7 km west of the outfall, at which point the fecal coliform count was 146/100 ml. The average non-plume fecal coliform concentration was <6.7/100 ml. Pathogens isolated included *Salmonella* and *Vibrio* spp., and a *V. cholerae* non-01 serovar, which was isolated from a nearby river also impacted by the sewage plant.

Clearly, the risk of being exposed to *Salmonella* and *Shigella* and to viruses, e.g., rotavirus, enteroviruses, and hepatitis viruses, can be significant, since a variety of organisms in sewage can cause infections in divers, notably enteritis.

In studies in Chesapeake Bay, including sewer outfalls, the numbers of bacteria have been shown to be significantly higher than in clean Chesapeake Bay water. Furthermore, a study done at the Anacostia River in Washington, D.C., revealed a significant seasonality in bacterial population size. Thus, in the summer months, a greater probability of infection may exist, at least for certain types of bacteria, both aerobic and anaerobic.

Potentially pathogenic anaerobes can be isolated from anoxic, polluted water. Several of these anaerobic bacteria can produce enterotoxins. Potentially, a larger number of anaerobes can be found in polluted waters, many of which can colonize the gear and skin of divers. These findings are summarized elsewhere in this volume.

Finally, what may appear to be safe seawater diving sites, when tested by conventional tests, may be hazardous because many organisms exposed to seawater do not die-off, as previously believed. Instead, they may enter a survival mode (5,6). Coliforms and pathogens may persist, even though they cannot be cultured, and they may remain potentially pathogenic. The reality, revealed by techniques of direct microscopy, is that no change occurs in the total number of bacteria detected by microscopic methods, but the ability to culture these bacteria decreases the longer the bacteria are exposed to seawater.

These findings place a new perspective on problems of diving in polluted water. They suggest that the oceans, where we have been dumping domestic and industrial wastes for decades, may be accumulating a host of viruses and bacterial pathogens. These may represent a hazard that may be likened to a ticking time bomb. We are only now beginning to realize that seawater of high salinity, low temperature, and low nutrient concentration may harbor a reservoir of pathogens in a dormant, but viable, state.

For the commercial diver repairing equipment in a sewer, organisms may be inhaled, skin abrasions or wounds may be infected, or bacteria may accumulate on diving gear, since bacteria rapidly colonize surfaces of gear exposed to polluted water. The papers presented in this section describe in greater detail the kind of microbiological risks which may be facing society as increasingly our coastal waters and estuaries and the open oceans are used for waste dumping. This conference lifts the dialogue on microbiological hazards and risks for divers in polluted waters to yet another level, providing new knowledge and the hope that the information provided will ultimately provide the basis for diver protection.

References

1. Singleton, F.L., J.W. Deming, E.R. Peele, B.Z. Cavari, B.A. Gunn and R.R. Colwell. 1983. Microbial communities in surface waters at the Puerto Rico dumpsite. Wastes in the Ocean V-1: Industrial and Sewage Wastes in the Ocean. John Wiley & Sons, Inc.

2. Grimes, D.J., F.L. Singleton, J. Stemmler, L.M. Palmer, P.R. Brayton and R.R. Colwell. 1984. Microbiological Effects of Wastewater Effluent Discharge into Coastal Waters of Puerto Rico. Water Res. 18 (5): 613-619.

3. Baya, A.M., P.R. Brayton, V.L. Brown, D.J. Grimes, E. Russek-Cohen and R.R. Colwell. 1986. Coincident plasmids and antimicrobial resistance in marine bacteria isolated from polluted and nonpolluted Atlantic samples. Appl. Environ. Microbiol. 51:1285-1292.

4. Colwell, R.R., P.R. Brayton, D.J. Grimes, D.R. Roszak, S.A. Huq, and L.M. Palmer. 1985. Viable, but non-culturable *Vibrio cholerae* and related pathogens in the environment: implications for release of genetically engineered microorganisms. Bio/Technology 3:817-820.

5. Roszak, D.B., S.A. Huq, P.R. Brayton, L.M. Palmer, D.M. Rollins, D.J. Grimes and R.R. Colwell. 1984. "Survival, viability and virulence studies of selected pathogens in aquatic systems of potential hazard to divers." NOAA Undersea Research Symp., May 22-24, 1984.

6. Roszak, D.B. and Colwell, R.R., 1987. Survival strategies of bacteria in the natural environment. Microbiol. Rev. 51:365-379.

Diving Equipment as Protection Against Microbial Hazards

DAVID M. ROLLINS
Naval Medical Research Institute
Infectious Disease Department
Bethesda, Maryland

Introduction

The Naval Medical Research Institute and the University of Maryland joined forces to investigate the potential presence of microbiological hazards in operational diving areas. The objectives were to identify pathogens in the waters of New York Bight, the Atlantic Marine Center in Norfolk, the Pacific Marine Center in Seattle, and the Anacostia River, to evaluate the degree of contamination before the divers went into the water and after they came out, and then to evaluate the effectiveness of the diving equipment in protecting divers against the pathogens.

Diving Equipment

A diver who is fully exposed to the water is at risk from any pathogens that it may harbor. The traditional wet suit and SCUBA gear offer scant protection against microbiological hazards. Even equipment that is appropriate for polluted water, such as a variable-volume suit and a helmet, offers many points of entry for contaminated water. Even when protective equipment is used properly, the chances of seals breaking and water getting into the dry suit are fairly good.

When diving in polluted water, it is a good idea to use a helmet that can be readily sealed to the suit, such as the Kirby Morgan-style helmet, which covers the entire face, and the Agga mask, which offers better peripheral vision and sits a little closer to the face.

Contamination of the Diving Equipment

Bacteria are most likely to colonize the interior parts of the suit: around the respirator, the regulator area of the valves, and around the seals of the helmet. Studies done by Itzac Grook and colleagues at the Naval Medical Research Institute showed that when a diver's ears are occluded with a hood and he sits for 25 to 30 minutes, the bacterial concentration in the ear rises dramatically, regardless of whether the diver has been in the water or out of the water. The bacteria in this particular test are believed to be type that normally occurs in the ear, but their dramatic increase was most likely a response to increased humidity and temperature. This problem can be alleviated by pretreating the suit with a dilute acetic acid solution.

A body of polluted water basically falls into one of three general categories: clean, moderately polluted, and heavily contaminated. For our studies, we employed a variety of suit and helmet types. The Kirby Morgan Mark I helmet or a Superlight 17 helmet can be tethered to an air supply at the surface, and it offers the diver complete protection. All parts of the suit, including the gloves and boots, are sealed off, sometimes even double-sealed, to produce a dry environment for the diver.

We evaluated the various suits and masks by taking microbiological samples from the surfaces of the suits and cultured them in the appropriate media. We used four different divers, for whom we cultured samples from the ear, nose, throat, and helmet. We examined the ports on the suits for external leakage, both before and after the divers entered the water. We checked the seals and ports and various sites on the suits, boots, and gloves and inside of the helmet to determine whether anything had leaked through and, if so, whether or not it had had a residual effect on the helmet.

The suits typically displayed similar levels of bacterial contamination. We had the four divers conduct two dives each, one in the morning and one in the afternoon, wearing the same helmet each time. By the afternoon it was clear that even before the second dive, the bacteria that had colonized from the morning dive were still present and their numbers had actually grown. Those bacteria levels were elevated even further after the second dive. These data proved that bacteria were present and that the diver was bringing them up with him. We found that *Aeromonas* bacteria in particular were colonizing the helmets.

Interestingly, many bacteria in the aquatic environment possess structures that are ideal for adherent colonization of surfaces. These particular bacteria had coalesced on the external surface of the helmets. But curiously, some strains do not possess the ability to adhere to surfaces, in this case, the masks, suits, and divers themselves.

When bacteria adhere to a surface, they produce a biofilm that eventually builds up. The biofilm can get so thick it has been known to slow down boats. If the biofilm is not removed, it causes a dry suit to degrade much more rapidly than it normally would. Bacteria can be removed from the suits with disinfectant procedures.

Some dry suits feature a subsurface layer with air pockets, which causes them to be buoyant. If such a suit develops cracks, the surface is broken through to the subsurface layer, onto which bacteria may lodge. But the surfaces of suits vary quite a bit. We looked at three different suits in our study: two smooth-surface suits, the Viking and Agualla, and one textured suit, the Poseidon unisuit. We found that the cracks and crevices of some of the textured suits are a haven for the proliferation of bacteria and various other debris.

Exposure Risks

What typically happens during operations in polluted water is that the diver is protected from the water when going in, but upon coming out has to break the seals and is thus exposed to pathogens. In addition, the tenders who deal with the diver are exposed to whatever the diver was exposed to at the bottom. Such topside personnel typically are unaware that they are at risk.

The diver is also at risk whenever he dons or sheds his suit without disinfecting it. So we considered a variety of disinfecting techniques and found a number of disinfectant sprays that significantly reduce the bacterial contamination of the equipment. Such sprays avoid the need to scrub the suits, which wears them down. When the divers surfaced, we sprayed the suits with Amway or zepamene or betadyne and let them stand for 10 minutes. Then we washed the suits with water, which significantly reduced the number of bacterial isolates that we could obtain from the suits' surfaces.

Conclusion

Much remains to be discovered about protecting divers from the hazards of polluted water. While certain techniques and precautionary measures can reduce the risk of contamination to both divers and topside personnel, old habits die hard, and it may take a strong educational effort to change current procedures for diving in contaminated water.

Monitoring and Significance of Aquatic and Terrestrial Bacteria in the Aquatic Environments of Divers

SAM W. JOSEPH
Department of Microbiology
University of Maryland, College Park, Maryland

J. B. CONWAY
Graduate School of Public Health
San Diego State University, San Diego, California

S. G. KALICHMAN
Consulting Public Health Engineer, San Diego, California

Introduction

Public health studies were conducted to assess the health risk resulting from the City of San Diego's wastewater discharge to those persons who dive near the Point Loma kelp bed or consume seafood taken from the bed. Bacteriological quality and conformance with bacteriological standards of kelp bed waters for the period April 1, 1985, to August 31, 1986, were reviewed. The diver health risk study made between June 18 and September 1, 1986, was a descriptive study of the health risk to divers using the Point Loma kelp bed. Of the various users of the kelp bed, the divers were most at risk because they came into contact with waters containing the highest bacterial indicator levels. Although high levels of total and fecal coliform organisms were frequently found at the outer edge of the kelp bed and were often found inside the bed, the diver studies made in 1986 and in an abbreviated study in 1985 indicated that the health risk to divers was low.

It appears that the aquatic environment is contaminated with a mixture of avirulent, virulent, and opportunistically pathogenic microorganisms. The list of pathogenic and potentially pathogenic organisms is quite lengthy and includes bacteria (both gram-negative and gram-positive), fungi, parasites, and viruses. Potential bacterial pathogens in the environment include terrestrial types, which represent external contamination, fecal or otherwise, and autochthonous types, which are commonly present in the water and under certain circumstances are capable of causing infection. While there are standards available to regulate contamination of terrestrial types, it is virtually impossible to establish regulations for microorganisms which are normal aquatic residents.

This report discusses two separate studies involving terrestrial and autochthonous types of organisms. The first study involving an autochthonous bacterium was performed by teams from the Naval Medical Research Institute and the University of Maryland in the Chesapeake Bay area while monitoring a diver training session in the Anacostia River near Washington, D.C. The second study was performed in San Diego Bay near Point Loma, the site of a sewage outfall, and involved monitoring of a commercial and sports diving area for terrestrial organisms.

Results

Chesapeake Study

Studies began in the Anacostia River near the Naval School of Diving and Salvage (Figure 1). The site was visibly contaminated at the time of the study and had a temperature range of 28°C in August to 0.5°C in February. Samplings of surface water, bottom water, and sediment revealed a total viable count ranging from approximately 6.5×10^7 colony forming units (CFUs) in sediment to about 10^2 CFU in surface water. Of these total numbers of organisms, a small percentage of anaerobic

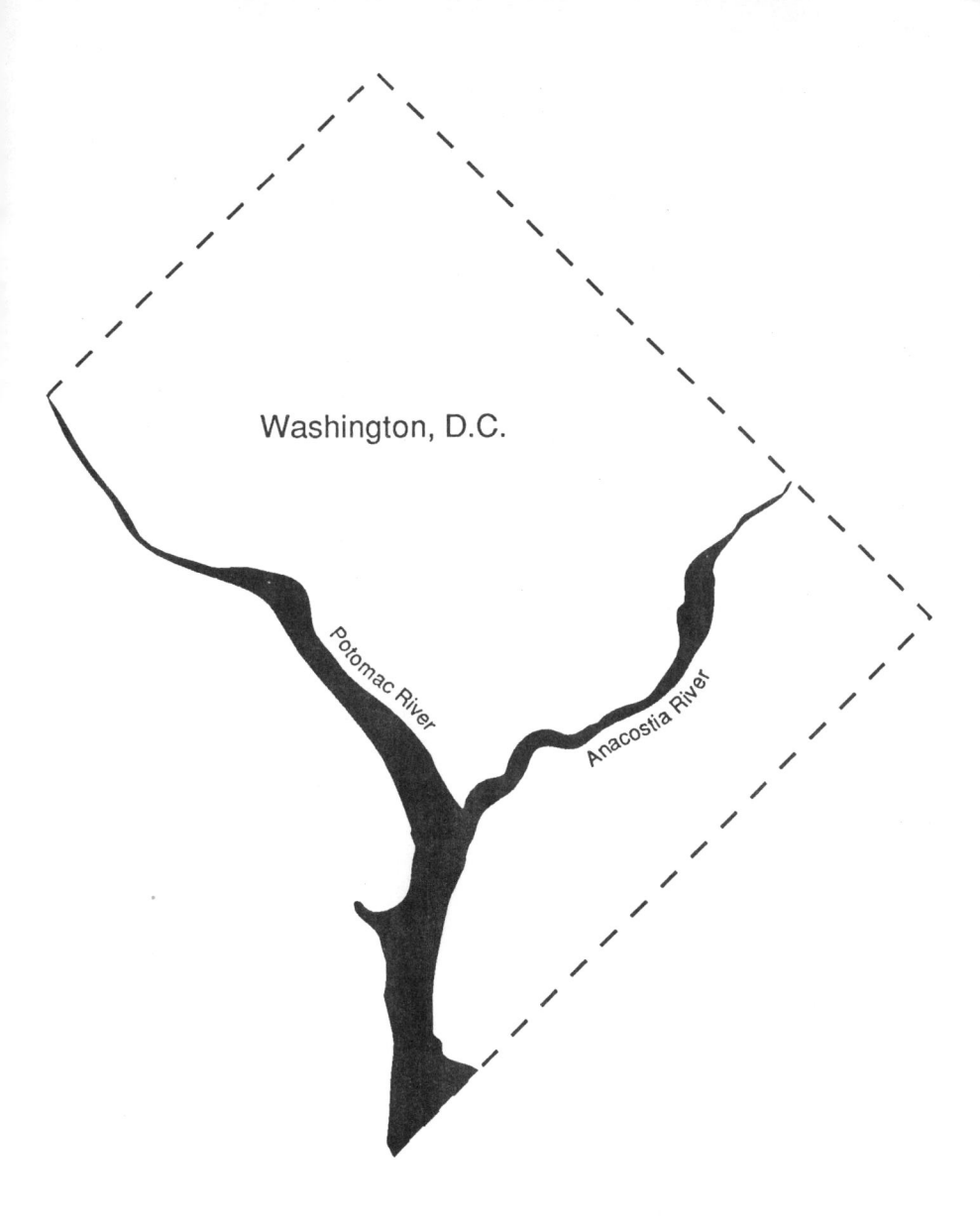

Figure 1. Map of Washington, D.C., showing locations of Potomac and Anacostia rivers.

organisms was found, which was surprising at the time because it was presumed that sites being sampled were mostly aerobic environments. This has even greater significance now because there has been a recent report by Kay and colleagues (1) that some types of anaerobic bacteria may be capable of causing intestinal disease in humans. Studies on toxicity of anaerobes isolated from the New York Bight area confirmed this suspicion (2).

While a large number of potentially pathogenic bacteria were isolated from the site, a single incident focused attention on the aquatic-born bacterium, *Aeromonas* spp. During the course of the studies, a diver suffered a primary *Aeromonas* infection of the soft tissue of the leg after a puncture wound had been implicated. Subsequent cultures showed this to be a dual infection actually caused by two species of *Aeromonas*. Initial identification of these two strains as *Aeromonas hydrophila* and *A. sobria* (3) has been updated using more recent methodology and information, and they are recognized now as arabinose-positive *A. hydrophila* and *A. jandaniensis*, respectively (4). At the time, *Aeromonas* was considered to be an opportunistic pathogen, causing disease only in people who were immuno-compromised--a common theme in almost all of the scientific articles published before 1975. However, the diver in this situation was a normal, healthy individual from whom two separate species of *Aeromonas* had been isolated, an event which had not been published previously. In fact, one of the two species, *A. jandaniensis*, had never been published as a human isolate. The overwhelming question at the time was whether this was an incidental situation caused by a randomly occurring *Aeromonas*, or whether there were large enough numbers present to cause concern that this organism was a serious potential pathogen.

Therefore, surface water and sediment were sampled and studied with the finding that *Aeromonas* constituted a very significant proportion of the total viable bacterial counts of the samples taken from the Anacostia River. In fact, the *Aeromonas* count was even higher than the total coliform and the fecal coliform counts observed at that time (5). These observations have been verified subsequently by a number of other studies which showed that *Aeromonas* counts seemed to rise in polluted (not necessarily by fecal matter) waters.

A subsequent study (6) performed in the Chesapeake Bay by a team of investigators from the University of Maryland and from the Naval Medical Research Institute surveyed many sampling stations from Baltimore Harbor to the mouth of the Chesapeake Bay, representing salinity ranges of 0 to 17 ppt. The findings showed the counts to be as high as 5×10^3 CFUs, which occurred most frequently in water having >15 ppt salinity. There was a seasonal distribution with the largest numbers found during the summer months. Seasonality was also observed in Australia by Burke et al. (7), who noted in a clinical study that when the water temperature was higher, the *Aeromonas* counts were higher and, consequently, more *Aeromonas* infections were seen in patients.

Having ascertained that large numbers of *Aeromonas* were present in the environment, investigators made an attempt to determine the ability of these organisms to adhere to divers and their equipment--adherence being the first step in the infection process. By measuring the presence and absence of these organisms in 15 divers before and after a 30-minute swim, it was found that aeromonads significantly adhered to the masks, nasal passages, ears, and throats of divers (Table 1). Although numbers were lower in studies conducted during the colder months (water temperature, 13°C in October vs. 28°C in August), the difference in bacterial numbers before and after the swim was still significant.

These studies were followed by a major investigation to determine the pathogenic nature of *Aeromonas* (8). Both clinical and environmental *Aeromonas* were studied, with the conclusion that not all of the aeromonads in the aquatic environment were potentially pathogenic for humans. On the other hand, there seemed to be a subset of aeromonads in the environment which were found in greater numbers in *Aeromonas*-associated human disease.

Table 1. Incidence of Aeromonas spp. recovered from divers and their diving equipment before and after swimming in the Anacostia River.

Month	No. of Aeromonas cells per ml[a]	Water temp (°C)[b]	Before swim				After swim			
			Mask	Ear	Nose	Throat	Mask	Ear	Nose	Throat
Aug	300	28	0/15	0/15	0/15	3/15	13/15	14/15	1/15	4/15
Oct	59	13	0/10	0/10	0/10	0/10	2/10	9/10	0/10	2/10

[a]Count in Anacostia River water at time of sampling.
[b]Temperature of Anacostia River water at time of sampling. Adapted from Siedler et al. 1980. Appl. Environ. Microbiol. 39:1010-1018.

One question which remained unanswered involved duration of colonization. Because of the nature of the population being studied, it was difficult to maintain follow-up to determine the length of time that these *Aeromonas* counts persisted.

San Diego Study

This study was conducted to assess the health risk resulting from the City of San Diego's wastewater discharge to those persons who dove near the Point Loma kelp bed or consumed seafood taken from the bed. Of the various users of the kelp bed, the divers were most at risk because they came into contact with waters containing the highest bacterial indicator level (9).

The study was performed from May through September of 1986 around the kelp bed off Point Loma (Figure 2). The kelp bed is about five miles long and one mile wide. There is an outfall pipeline reaching from the sewage disposal plant on Point Loma to an area approximately three miles offshore and one and a half miles beyond the outer edge of the bed. Although the coordinates of the stations did not change, the boundaries of the kelp bed did vary over extended periods.

Total coliforms, fecal coliforms, and enterococci were measured in samples taken from 14 stations identified in Fig. 2 at four to five different depths as well as from surface and bottom water. Enterococci were monitored for general information and to compare with coliform values.

The studies of Cabelli et al. (10) equate the presence of enterococci with the potential for causing highly credible gastrointestinal (HCGI) symptoms. The Environmental Protection Agency (EPA) proposal essentially required that recreational waters show fewer than 35 enterococci/100 ml of water. This was in contrast to the previous coliform-based requirements, i.e., <1000 total coliforms/100 ml and <200 fecal coliforms/100 ml of water.

The total number of divers recruited during this study was 346 individuals, almost a ten-fold increase over the number employed in a very abbreviated study performed in 1985. The major reason for this increase was the recruiting of divers on site by boat in the vicinity of the Point Loma kelp bed. The number of new divers contacted by this method was 316 out of the 346; the remaining 30 divers came from individuals who participated in 1985.

Divers were recruited from dive sites throughout the kelp bed, although divers from sites along the outer edge were of most interest. A second, smaller concentration of divers was located offshore from Point Loma College. During the study, divers were contacted where they were found and, on some sampling dates, most diving boats were approached and every diver aboard was recruited. The highest single number of divers in any one grid was ten, which occurred within the kelp bed just south of the outfall. The concentration of divers in these two locations might have been influenced by the existing kelp canopy or knowledge of the location of scallops, abalone, or lobster.

A profile of these 316 new divers, who made 1,371 dives during the study, is given in Table 2. These divers were mostly male (280 versus 36 females), averaged 33.5 years of age, had nine years

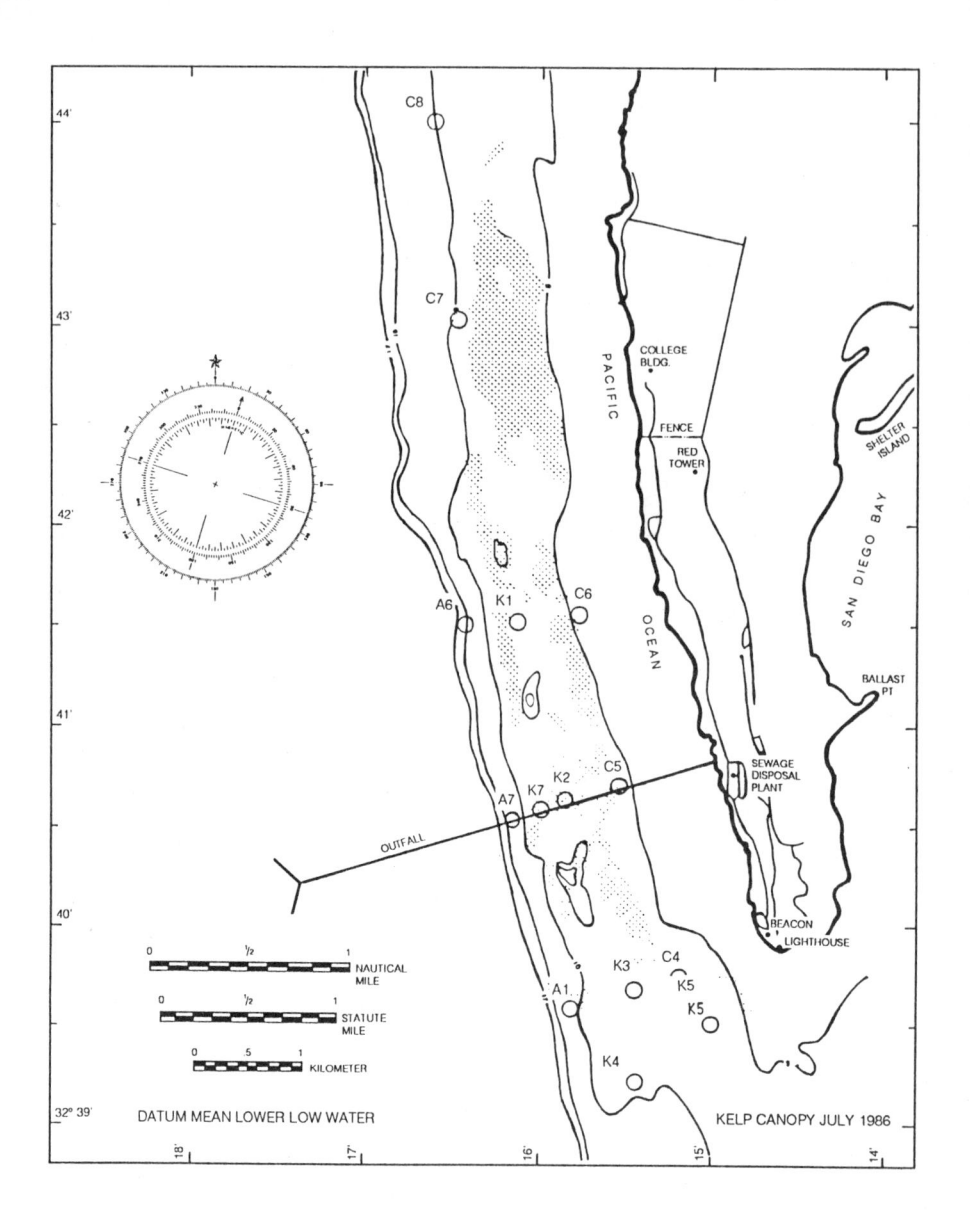

Figure 2. Point Loma kelp bed sampling stations.

of diving experience, were predominately recreational divers (23 were involved in research and 20 did some commercial work), and made on the average five dives in the Point Loma kelp bed area per month. The number of divers who reported removing their masks or mouthpiece was 223; getting water in their mouths, 240; swallowing water, 146 (the average amount swallowed, 3 tablespoons, approximately 44 ml^3. Fifty-seven divers said that they ate raw seafood underwater. The average depth of a dive at Point Loma was 52 feet, the average time in the water was 42 minutes, and usually two dives were made per day. Two hundred and eighty-six of the divers reported taking seafood from the vicinity of the Point Loma kelp bed, 281 ate what they caught, and 212 cooked the food before they ate it.

Twelve illnesses which fit the HCGI symptoms, as defined in EPA epidemiological studies, were reported. If all HCGI illnesses reported were diving associated, then this represented 8 HCGI cases per 1,000 dives. The EPA water-contact criteria, which called for the use of *Enterococcus* as the indicator organism for marine waters, recommended a maximum allowable geometric mean enterococcus concentration, which would permit an estimated 19 illnesses per 1,000 swimmers. Only kelp bed stations A-6, A-7, and C-7 at the 40- and 50-foot depths exceeded the EPA geometric mean enterococcus criteria. All other stations met the criteria. Inside the bed, at one or more depths, the coliform standards were never or rarely met. Surface waters always conformed to the standards, except at one station for a short period of time (Tables 3-5).

Bacteriological samples were collected in the vicinity of the outfall at four to five different depths on six different dates and from 21 different divers during 14 dives. State bacteriological standards were exceeded during four of five dives. However, none of the divers recruited for this part of the study reported any illness.

Table 2. Profile of divers recruited by the student diver team at the Point Loma kelp bed.

Number of divers	316	
Male to female ratio	280:36	
Mean age (years)	33.5	(range 16-68)
Mean diving experience (years)	9	(range 1/12-40)
Divers with 1 year or less experience	56	
Average divers per day	2	(range 1-5)
Average divers per month	5	(range 1-30)
Mean diving depth (ft)	52	(range 1-161)
Mean diving time (min)	42	(range 5-98)
Divers involved in research	23	
Divers involved in commercial work	20	
Divers removing mask and/or mouthpiece	223	
Divers getting water in mouth	240	
Divers swallowing water	146	
Mean volume of water ingested in cubic centimeters[1]	44	(range 15-237)
Divers taking seafood at Point Loma	286	
Divers eating raw seafood under water	57	
Divers who eat what they catch	281	
Divers who cook the seafood before eating	212	
Divers who reported being ill	62	
Highly credible gasterointestinal illnesses (HCGI)	4	

[1] Volume based on rough estimates in tablespoons reported by the divers.

Table 3. Percent total coliform samples greater than 1000/100ml.

Station		Depths							Entire
		TOP	10'	20'	30'	35'	40'	50'	60'	Station
Outer edge of Kelp bed	A1	0	6	11	-	-	22	33	44	17
	A6	0	11	33	-	-	67	72	78	40
	A7	0	17	33	-	-	44	61	78	35
	C7	0	0	6	-	-	50	72	67	29
	C8	0	0	6	-	-	33	56	44	21
Kelp bed	K1	0	0	0	55	-	-	-	-	15
	K2	0	0	20	-	36	33	-	-	18
	K3	0	0	0	-	30	11	-	-	8
	K4	0	0	0	-	-	33	22	44	17
	K5	0	0	0	-	0	22	-	-	4
	K7	0	0	18	-	50	44	-	-	22
Inner edge of Kelp bed	C4	5	0	0	0	-	-	-	-	2
	C5	5	0	0	0	-	-	-	-	2
	C6	0	0	0	0	-	-	-	-	0

Table 4. Percent total coliform samples greater than 10,000/100ml.

Station		Depths							Entire
		TOP	10'	20'	30'	35'	40'	50'	60'	Station
Outer edge of Kelp bed	A1	0	0	6	-	-	0	0	0	1
	A6	0	0	6	-	-	22	28	33	13
	A7	0	0	17	-	-	17	17	0	9
	C7	0	0	0	-	-	0	11	11	3
	C8	0	0	0	-	-	0	11	11	3
Kelp bed	K1	0	0	0	0	-	-	-	-	0
	K2	0	0	0	-	0	0	-	-	0
	K3	0	0	0	-	0	0	-	-	0
	K4	0	0	0	-	-	0	11	11	4
	K5	0	0	0	-	0	0	-	-	0
	K7	0	0	0	-	0	0	-	-	0
Inner edge of Kelp bed	C4	0	0	0	0	-	-	-	-	0
	C5	0	0	0	0	-	-	-	-	0
	C6	0	0	0	0	-	-	-	-	0

Table 5. Percent fecal coliform samples greater than 400/100ml.

Station		TOP	10'	20'	30'	35'	40'	50'	60'	Entire Station
Outer edge of Kelp bed	A1	0	6	6	-	-	28	28	33	15
	A6	0	11	17	-	-	61	67	56	33
	A7	0	6	22	-	-	28	50	56	24
	C7	0	0	6	-	-	44	67	56	26
	C8	0	0	0	-	-	17	22	44	11
Kelp bed	K1	0	0	0	55	-	-	-	-	15
	K2	0	0	0	-	27	22	-	-	10
	K3	0	0	0	-	0	0	-	-	0
	K4	0	0	0	-	-	22	11	33	11
	K5	0	0	0	-	0	11	-	-	2
	K7	0	11	9	-	20	33	-	-	14
Inner edge of Kelp bed	C4	5	0	0	0	-	-	-	-	2
	C5	0	0	0	0	-	-	-	-	0
	C6	0	0	6	0	-	-	-	-	2

An analysis of the data obtained in this descriptive study led to the following observations. Of the various users of the kelp bed (divers, seafood consumers, and surface water users), divers were most at risk because they came into contact with waters containing the highest indicator levels. Despite this apparent risk, demonstrated by high levels of total and fecal coliform organisms frequently found at the outer edge of the kelp bed and often inside the bed, the health risk to divers is low. There is, in general, a low reported incidence of infection in divers. Overall, the risk to seafood consumers and surface water users in this study did not appear to be significant.

Discussion

Divers

A review of the literature reveals little reliable information on the hazards of diving in polluted waters (11). There has been an apparent low incidence of reported infectious disease in divers; however, the potential for infection is present, as shown in a New York City outbreak which took place during the summer of 1982 (12). Gastrointestinal symptoms developed in 17 of 40 New York City Fire Department scuba divers following training dives in polluted coastal waters. Eight cases of parasitic infection were found in the 17 affected divers. This is the first report of enteric parasitic infection associated with scuba diving in sewage-contaminated coastal waters. The episode took place in the Hudson River, which is one of two rivers receiving more than 200 million gallons of raw sewage daily from New York City. The bacterial contamination in the Hudson River diving sites as measured by the total and fecal coliform counts ranged from 50 to 100 times greater than the New York City standard for bathing water. Except for this incident, most of the medical data on divers are essentially anecdotal. In general, the most common types of infections associated with polluted waters are ear infections and skin rashes followed by mild to severe gastrointestinal disorders lasting from 2 to 10 days (13). Wound infections and respiratory infections are less common, but do occur. Some investigators have noted a significantly higher incidence of infection of cuts on the hands of divers after diving in polluted waters (14). The amount of health risk to a diver depends on the degree of pollution of the water in which the diver is immersed, the amount of sea water the diver may drink, the protection provided by the particular suit and hood worn, and the health of the diver. The risk is increased in more polluted waters and when more water is consumed.

Studies have shown that diving equipment can acquire significant quantities of bacteria during diving in polluted waters, and that disinfection by one of several means can effectively reduce much of the contamination (15). Relatively simple disinfection techniques combined with prophylactic use of ear drops can protect the divers and their support personnel from contact with high levels of potentially pathogenic bacteria. Proper cleaning between uses is recommended. The wet suits used by scuba divers allow the surrounding waters to come into direct contact with the skin of the divers, thereby presenting a potential for infection.

The low reported incidence of infection in divers may be due to the age range and physical condition of the typical diver, factors which may provide increased resistance, making them less susceptible than the average individual to an attack by pathogens.

Surface Water Users

Two apparent bacterial outbreaks, one in persons participating in a snorkel swimming event and the other in persons windsurfing in sewage-polluted waters, have been reported. In the snorkel-swimming event, which took place in 1983 at the Bristol City Docks in England, 27% (21/77) of the swimmers experienced gastrointestinal symptoms (16). The HCGI symptoms of Cabelli were met in 17 of the 21 swimmers. In Great Britain, the European Economic Community (ECC) bathing guidelines recommend that bacteriological levels not exceed 10,000 total coliforms and 2000 *Escherichia coli* per 100 ml, and no viruses should be present. The ECC's bacteriological guidelines were exceeded during the event. (One sample was run for viruses and found to be negative.) However, a similar proportion of swimmers reported gastrointestinal symptoms in an equivalent 1982 event when the water quality was within EEC guidelines.

The second reported outbreak took place at Quebec City during a Windsurfer Western Hemisphere Championship in 1984 (17). In the bay used for the windsurfing event, fecal coliform and enterococci concentrations were frequently above acceptable limits. Seventy-nine windsurfers and 41 controls were studied over a nine-day period for occurrence of symptoms of gastroenteritis, otitis, conjunctivitis, and skin infection. Comparing the two groups, relative risks were found to be 2.9 for the occurrence of one or more of these symptoms and 5.5 for symptoms of gastroenteritis. The relative risk for each health outcome was measured as the ratio of the cumulative incidence of symptoms among windsurfers to that among the controls. The relative risk increased with the reported number of falls in the water. The HCGI symptom categorization was not used, but illness must not have been too severe because no one withdrew from an event for health reasons.

Conclusions

There is an overall dearth of information regarding infectious disease incidence in diving populations. There is only sporadic monitoring of diver health and very few reports of designed surveillance studies. This topic usually receives attention only when there is an eventful outbreak involving a population of divers in some newsworthy activity.

Because divers are frequently exposed to contaminated aquatic environments, our increasing knowledge of the potential pathogens in such environments causes us to believe that we should become more attentive in measuring diver health.

A system for gathering quantitative data in a standard manner should be established to enable us to measure not only the health trends, but the role of specific agents in contributing to health problems. It would certainly be of interest to see how the standard morbidity ratio of these "sea-goers" compares with the average population of "land-lubbers."

Acknowledgments

Many persons assisted in the San Diego study, and their advice and contributions are greatly appreciated. We especially wish to thank the members of the Health Risks Advisory Committee and the Designation Study Team; the many divers who participated in the study; Ladin Kelaney and Mike McCann, Executive Officer and Senior Engineer, respectively, California

Regional Water Quality Board, San Diego Regional; Dr. Alfred DuFour, Health Effects Research Laboratory, USEPA; James Kreissl, Program Manager, Innovative and Alternative Technology Staff, Wastewater Research Division, USEPA; Dr. Otis Sproul, Dean, College of Engineering and Physical Sciences, University of New Hampshire; Bill Jopling, California Department of Health Services; and representatives of the Kelco Company, Scripps Institution of Oceanography, Los Angeles County Sanitation Districts, the U.S. Navy, and the City of San Diego.

The Chesapeake study was funded in part under Naval Medical Research and Development Command Work Unit nos. MR04105.01.0049 and MR04105.01.0050 and National Oceanic and Atmospheric Administration (NOAA) contract 01-8-MI-2027 and NOAA grant 04-8-M01-71.

References

1. Kay, B.A., T. Rahman, D.A. Sack, M.A.K. Azad, K.A. Chowdhury, and R.B. Sack. 1987. Pathogenic *Bacteroides fragilis* as a cause of human diarrheal disease in Bangladesh. In: Proceedings of the 23rd Joint M.S.-Japan Cholera Conf. U.S.-Japan Coop. Med. Sci. Prog. (In Press).

2. Daily, O.P., S.W. Joseph, J.D. Gillmore, R.R. Colwell, R.J. Seidler. 1981. Identification distribution and toxigenicity of obligate anaerobes in polluted waters. Appl. Environ. Microbiol. 41:1074.

3. Joseph, S.W., O.P. Daily, W.S. Hunt, R.J. Seidler, D.A. Allen, R.R. Colwell. 1979. *Aeromonas* primary wound infection of a diver in polluted waters. J. Clin. Microbiol. 10:46.

4. Joseph, S.W., A.M. Carnahan, D.M. Rollins, R.I. Walker. 1989. *Aeromonas* and *Plesiomonas* in the environment: value of differential biotyping of Aeromonads. J. Diarrhoeal Dis. Res. 6:80.

5. Seidler, R.J., D.A. Allen, H. Lockman, R.R. Colwell, S.W. Joseph, O.P. Daily. 1980. Isolation, enumeration and characterization of *Aeromonas* from polluted waters encountered in diving operations. Appl. Environ. Microbiol. 39:1010-1018.

6. Kaper, J.B., H. Lockman, R.R. Colwell, S.W. Joseph. 1981. *Aeromonas hydrophila*: ecology and toxigenicity of isolates from an estuary. J. Appl. Bacteriol. 50:259.

7. Burke, V.J., M. Robinson, D. Gracey, D. Peterson, K. Partridge. 1984. Isolation of *Aeromonas hydrophila* from a metropolitan water supply: seasonal correlation with clinical isolation. Appl. Environ. Microbiol. 48:361.

8. Daily, O.P., S.W. Joseph, J.C. Coolbaugh, et al. 1981. Association of *Aeromonas sobria* with human infection. J. Clin. Microbiol. 13:769.

9. Conway, J.B., S.G. Kalichman. 1986. Health risk study of the Point Loma kelp bed. In: Request for Revision of Water Quality Objectives and Discharge Requirements: Point Loma Facility, San Diego, California.

10. Cabelli, V.J., et al. 1979. Relationship of microbial indicators to health effects of marine bathing beaches. Am. J. Public Health. 69:690.

11. Gottlieb, S.F. 1982. Assessing the potential of microbial hazards of diving in polluted waters. Marine Technol. Soc. J. 15:3.

12. Jones, C.J., A. Goodman, T. Cox, S. Friedman, S. Schultz. 1985. Scuba diving in polluted coastal waters. Oceans 85: 950-961.

13. Daily, O.P., J.C. Coolbaugh. 1985. Infectious disease hazards associated with polluted waters. Oceans 85: 942-944.

14. Phoel, W.C. 1981. NOAA's requirements and capabilities for diving in polluted waters. In: Colwell R.R., ed. Workshop Report on Microbial Hazards of Diving in Polluted Waters. College Park, MD: Maryland Sea Grant Publ UM-SG-TS-82-01. Baltimore: University of Maryland, 9.

15. Coolbaugh, J.C., O.P. Daily. 1985. Protection of divers in biologically polluted waters. Oceans 85: 952-955.

16. Philipp, R., E.I. Evans, A.O. Hughes, S.K. Grisdale, R.G. Endicott, A.E. Jephcott. 1985. Health risks of snorkel swimming in untreated water. Int. J. Epidemiol. 14:624-627.

17. Dewailly, E., C. Poirier, F.M. Meyer. 1986. Health hazards associated with wind surfing on polluted water. Am. J. Public Health. 76:690.

Pathogenic *Vibrionaceae* in Patients and the Environment

MICHAEL T. KELLY
Microbiology Department
Metro-McNair Clinical Laboratories
Vancouver, British Columbia V57 1B5

Introduction

Two general categories of pathogenic bacteria may be found in marine and/or freshwater environments: (a) those introduced into the water primarily from terrestrial sources, such as members of the family *Enterobacteriaceae*, and (b) those which are natural inhabitants of aquatic environments (autochthonous), mainly members of the family *Vibrionaceae*. Diseases caused by autochthonous bacteria in the family *Vibrionaceae* include severe watery diarrhea due to *Vibrio cholerae* (1); seafood-related diarrhea as well as extraintestinal infections due to *V. parahaemolyticus* (2); and wound infections and lethal septicemia due to *V. vulnificus* (2). In addition, members of the *Aeromonas hydrophila* group and *Plesiomonas shigelloides* may be autochthonous human pathogens (3,4). The occurrence of autochthonous human pathogens in two different marine environments was assessed over a period of nine years.

Environmental Studies

Surface water samples were collected from 21 sites around Galveston Island, Texas, during 5 years of study, and from 13 sites on the southern coast of British Columbia during 4 years of study. One-liter samples of water were collected in sterile Whirl-Pak bags and processed by membrane filtration. Bacteria collected on the filters were cultured on the surface of agar media for the recovery of *Vibrio, Aeromonas,* or *Plesiomonas* (5,6). Measurements of temperature, salinity, pH, and dissolved oxygen were taken *in situ* at each sample collection using portable instruments. Oyster samples were also collected from some sites and dissected and cultured (7,8). Isolates were identified according to standard methods (2,9).

Clinical Studies

Clinical isolates of *Vibrionaceae* were collected from hospital laboratories in Galveston, Texas, and Vancouver, British Columbia, and from a laboratory serving outpatients in Vancouver. Clinical and epidemiologic information was obtained for each patient who yielded a *Vibrionaceae* isolate to assess the clinical significance of the isolate and to identify possible sources of the infecting organism.

Results and Discussion

Vibrio vulnificus

In Vancouver, one *V. vulnificus* wound infection was detected in a person who had been swimming in the ocean. Four *V. vulnificus* infections were encountered in Galveston. A person who had been found floating face down in the ocean off Galveston Island subsequently developed fatal pneumonia and septicemia due to *V. vulnificus* (10). A commercial fisherman developed fatal primary septicemia (11). A third person developed *V. vulnificus* endometritis after engaging in sexual intercourse while swimming in the ocean (12). The fourth case was a cancer patient who developed septicemia following a *V. vulnificus* infection of a hand wound exposed to seawater. The latter two patients responded to prompt antimicrobial therapy and recovered from their infections.

The infections in Galveston prompted environmental studies on the occurrence of *V. vulnificus*. Initial sampling in the summer indicated that the organism was a common inhabitant of estuarine environments around Galveston Island (10). More extensive studies demonstrated a marked seasonal variation in the occurrence of *V. vulnificus* (5). The organism was present during the summer when average water temperatures were above 20°C, was uncommon during the winter, was favored by salinities below 16 ppt (5), and was recovered from 70% of oysters collected during July when water temperatures were 28°C, but absent in oysters collected in March (water temperature 15°C) (7).

Studies in Vancouver confirmed these findings. *V. vulnificus* has been recovered from estuarine environments in British Columbia only during the summer and in protected environments where water temperatures exceed 20°C but was not detectable at other times of the year when water temperatures are lower. The organism was favored by relatively low salinity in British Columbia environments and was also recovered from 21% of oysters collected from the British Columbia coast during the summer, but no oysters collected during the winter yielded the organism (8). The British Columbia *V. vulnificus* isolates are biochemically identical to those from Galveston Island. The lower number of *V. vulnificus* infections in Vancouver may be a reflection of the brief time that the organism is present in detectable amounts in northern environments.

Vibrio cholerae

Three cases of *V. cholerae* diarrhea were detected in Galveston during the period of study, part of the ongoing endemic focus of cholera that has been identified on the Gulf Coast (1), and three cases have also been identified in Vancouver over the past 4 years. One of these patients had acquired cholera while traveling; the other two had extraintestinal infections that were locally acquired. Environmental studies in Texas revealed that *V. cholerae* was a common inhabitant of aquatic environments. Most of the isolates were of non-01 serotypes not associated with epidemic disease, but follow-up of one cholera case resulted in recovery of 01 strains from aquatic sources in the patient's environment (13). Non-01 *V. cholerae* strains have also been recovered from estuarine and freshwater environments in British Columbia with a pattern of isolation similar to that of *V. vulnificus* in that the organism was recovered only during the summer from warm water environments of relatively low salinity.

Vibrio parahaemolyticus

Twenty cases of *V. parahaemolyticus* infection have been encountered in Vancouver over the past 4 years. Of these, 13 were intestinal infections and 7 were extraintestinal infections. Clinical and epidemiologic information is available for 13 of these patients (6): Three acquired their infections while traveling, but 10 of the patients acquired the infections locally and all of these had a history of exposure to seawater, seafood or both. The extraintestinal infections involved wounds exposed to seawater or external otitis after swimming in the ocean. The intestinal infections ranged from mild to severe, and one patient had a significant weight loss resulting from severe diarrhea of 2-weeks duration.

V. parahaemolyticus was isolated from environmental water samples in significant numbers only in July, August, and September during the 4-year study, and the occurrence of the organism was again associated with warm, low-salinity water conditions. *V. parahaemolyticus* was also recovered from 37% of oysters but only during the summer (8). Locally acquired *V. parahaemolyticus* infections occurred only when the organism was present in significant numbers in the environment, and occurrence of the organism in patients and the environment was closely associated (6), suggesting that persons exposed to estuarine environments during the summer may be at risk for development of *V. parahaemolyticus* infections.

Aeromonas hydrophila Group

Aeromonas infections have been the most commonly encountered of those studied; during 3 years, 232 infections have been documented, including 14 extraintestinal infections and 218 cases of gastroenteritis. Although an environmental exposure history is not as common with *Aeromonas* infections as with the other organisms studied, several patients developed extraintestinal infections after exposure of wounds to aquatic sources. Several patients with intestinal infections also have a history of exposure to aquatic environments or raw shellfish.

Aeromonas has also been commonly encountered in freshwater and estuarine environments on the British Columbia coast. Between 45 and 84% of water samples collected from 10 sites in the Vancouver area have been positive for the organisms, and organisms of the *Aeromonas hydrophila* group have been recovered at all times of the year. Their occurrence is favored by low salinity; they are present in higher densities in warm water, but they do not disappear during the winter. *Aeromonas* was recovered from 9% of oysters cultured, and oysters collected both in the summer and winter yielded the organism (8).

Plesiomonas shigelloides

Fifty-five cases of *P. shigelloides* diarrhea have been detected during 3 years of study in Vancouver. Clinical and epidemiologic data were collected for the first 30 patients, and the findings suggest that this organism is a significant enteric pathogen that may be environmentally acquired (14). Many patients acquired their infections while traveling, but 29% had no history of recent travel outside British Columbia although these all had a history of consumption of seafoods or untreated water or both. Most of the patients had a dysentery-like illness often associated with severe abdominal pain, and antimicrobial therapy significantly shortened the duration of illness. Although *P. shigelloides* was isolated from environmental water samples less frequently than were other *Vibrionaceae*, it was detected in 10% of samples cultured for the organism during 1.5 years of study. The organism was also recovered from 2% of oysters cultured. Environmental studies have not been carried out long enough to assess the influence of temperature and salinity on the occurrence of *P. shigelloides*.

Implications for Divers

These studies indicate that potentially pathogenic autochthonous bacteria occur commonly both in subtropical and temperate estuarine environments and that human infections may be associated with the presence of these organisms. Warm water and low salinity influence the occurrence of the organisms, suggesting that particular caution should be exercised during the summer. External otitis is a recognized diving hazard, but other types of infections, especially wound and gastrointestinal, should be investigated. In this regard, salmon farms have the potential to increase the numbers of human pathogens in the environment through the introduction of increased nutrient supplies and fish feces; after starting work on a salmon farm, four divers reported diarrhea, and ear, eye, and skin lesions that were of probable infectious origin.

References

1. Kelly, M.T. 1986. Cholera: a world-wide perspective. Pediatr. Infect. Dis. 5:5101-5105.

2. Tison, D.L., M.T. Kelly. 1984. *Vibrio* species of medical importance. Diagn. Microbiol. Infect. Dis. 2:263-276.

3. Holmberg, S.D., W.S. Schell, G.R. Fanning. 1986. *Aeromonas* intestinal infections in the United States. Ann. Intern. Med. 105:683-689.

4. Holmberg, S.D., I.K. Wachsmuth, F.W. Hickman-Brenner. 1986. *Plesiomonas* enteric infections in the United States. Ann Intern Med. 105:690-694.

5. Kelly, M.T. 1982. Effect of temperature and salinity on *Vibrio (Beneckea) vulnificus* occurrence in a Gulf Coast environment. Appl. Environ. Microbiol. 44:820-824.

6. Kelly, M.T., E.M.D. Stroh. 1988. Temporal relationship of *Vibrio parahaemolyticus* in patients and the environment. J. Clin. Microbiol. 26:1754-1756.

7. Kelly, M.T., A. DiNuzzo. 1985. Uptake and clearance of *Vibrio vulnificus* from Gulf Coast oysters (*Crassostrea virginica*). Appl. Environ. Microbiol. 50:1548-1549.

8. Kelly, M.T., E.M.D. Stroh. 1988. Occurrence of *Vibrionaceae* in pacific oysters. Diagn. Microbiol. Infect. Dis. 9:1-5.

9. Farmer, J.J. Jr., F.W. Hickman-Brenner, M.T. Kelly. 1985. Vibrio. pp. 263-277. In: E.H. Lennette, A.Balows, W.J. Hausler, Jr., H. J. Shadomy, eds. Manual of Clinical Microbiology. American Society for Microbiology, Washington, D.C.

10. Kelly, M.T., D.M. Avery. 1980. Lactose positive *Vibrio* in seawater: a cause of pneumonia and septicemia in a drowning victim. J. Clin. Microbiol. 11:278-280.

11. Kelly, M.T., W.F. McCormick. 1981. Acute bacterial myositis caused by *Vibrio vulnificus*. J. Amer. Med. Assoc. 246:72-73.

12. Tison, D.L., M.T. Kelly. 1984. *Vibrio vulnificus* endometritis. J. Clin. Microbiol. 20:185-186.

13. Kelly, M.T., J.W. Peterson, H.E. Sarles, Jr., *et al.* 1982. Cholera on the Texas Gulf Coast. J. Amer. Med. Assoc. 247:1598-1599.

14. Kain, K., M.T. Kelly. 1989. *Plesiomonas shigelloides* diarrhea: clinical features, epidemiology and treatment. J. Clin. Microbiol. in press.

Protection of Recreational Divers against Water-borne Microbiological Hazards

B. F. MASTERSON
Biochemistry Department
University College
Belfield, Dublin 4, Ireland

Introduction

For recreational divers, the notion of health risk is offputting and detracts from the enjoyment of the recreational amenity. Also, aesthetic deterioration frequently accompanying microbial water pollution exacerbates the disdain, though as a counter point to this, even in the best aesthetic conditions with high underwater visibility, the water body may still be polluted microbiologically. Diving in polluted waters carries the risk of infection by pathogenic microorganisms; indeed pathogenic organisms have been recovered from divers exposed to polluted waters (1). Many engaged in diving activities will be aware of anecdotal cases, such as ear, nasal, and even bladder infections attributed to the baneful effects of water pollution. Concern among divers about the risk of water-borne virus infection, which is imagined as hugely insidious, has increased in recent times. It has been stated bluntly that water polluted with human sewage causes hepatitis-A (2). The nature and extent of hazards to health are of much concern to recreational divers in contact with polluted waters.

The Nature of the Hazard

Survey of the literature shows that there are but few clinical reports of outbreaks and cases of serious illness associated with use of recreational waters, although an unknown proportion of incidences of a minor nature may be unreported or their origins unidentified. Enteric infections may be associated with sewage pollution or contamination of the water by the shedding of faecal organisms (3) by the users themselves. Only a small number of typhoid outbreaks have been documented (4), all of which were linked to grossly polluted water unlikely to be encountered by recreational users. The few cases of poliovirus infection putatively associated with water were considered inconclusive (5), consistent with the findings of Moore (6). It is likely that a sole reported outbreak of hepatitis-A (7) among bathers at a fresh water lake was caused by inadvertent use of the lake water for drinking purposes. Two small outbreaks of shigellosis have been reported (8) for which no source of infective pollution was identified, and one further outbreak occurred at a stretch of river which was highly polluted (9). Evidence that a small outbreak of coxsackievirus enteritis involving five children (10) and a larger one with 33 young lake-side campers (11) were swimming-related has been criticized (5). An outbreak involving over 300 cases most likely induced by Norwalk virus, was attributed to recreational use of an artificial lake, but the source of pollution was not discovered (12). A high incidence of gastroenteritis was found among snorkle swimmers competing in polluted dockland waters (13).

The sources of respiratory, ear, eye and skin infections are most likely to be found among the water users themselves, save possibly where there is heavy pollution by sewage. Reports refer most frequently to swimming-pool incidences, where disinfection procedures were believed to be ineffective (14). A significant association, though, was reported (15) in children under 16 years of age between respiratory illness (and gastroenteritis) and swimming at freshwater beaches; it was unclear whether infection was due to sewage pollution, the extent of which may have been understated, or to the spread of infection between swimmers themselves. Otitis externa is the most commonly reported water contact illness (16, 17), and the most prominent causative organism is *Pseudomonas aeruginosa* (18). Transmission of the infection among commercial divers has been reported (19); with frequent swimmers it has been encountered more often in fresh water than in sea or swimming-pool water (20). Of special importance to divers is the observation (21) that *P. aeruginosa* may become established in benthic sediments, possibly following introduction of the organism by sewage discharges. Although an outbreak of swimmers'-itch has been reported (22), low incidences of skin infections are usually recorded in illness surveys among recreational water users. An association was found between the degree of water pollution and ear, eye and skin (and enteric) infections among windsurfers competing on waters judged unsafe for bathing (23).

Traumatic wounds incurred by divers may become infected by autochthonous pathogens (24, 25); incidences are rare, though there is a growing list of reports of *Aeromonas hydrophila* infection (26 - 31) and the organism is also associated with enteric disorders (32).

There is concern about leptospirosis, an infection associated with the use of fresh water contaminated with animal faecal matter, and about water-borne camphylobacters which cause diarrhoea; however, there are no reported cases associated with recreational water use for these pathogens. Generally then, with rare exceptions, clinical reports involve illnesses that are minor in nature and are usually associated with substantial water pollution. Certainly there is some degree of microbiological hazard to health involved in recreational diving; however small, it is important to know the extent of it.

The Extent of the Hazard

The orthodox approach to the assessment of hazard to health imposed by microbiological water pollution is to attempt to associate in a quantitative way the degree of pollution with the level of incidence of related illnesses. Even for pathogenic organisms for which convenient methods of enumeration are available the water densities are usually well below those measurable. The stratagem is to assess water quality by enumerating "indicator organisms" (33, 34) which accompany the pathogens; indicator organism densities measured by standard methods are used to index the levels of the accompanying pathogens. Measured indicator organism levels then, should correlate with the observed illness incidences as determined by epidemiological survey. Two small studies have been completed recently (13, 35); older and more elaborate studies (6, 36-42) carried out in various parts of the world have been critically reviewed elsewhere (43, 44).

Public health agencies have employed the correlations derived from such surveys to set regulatory limits for indicator organism densities within which "acceptable" levels of illness may be anticipated among recreational water users (45, 46). The European directive on bathing water quality (45) in part for instance, specifies fortnightly sampling at designated bathing sites during the months of main recreational use; for eighty percent of the sampling days, total and faecal coliform densities must be within certain limits. In reality microbial monitoring data are very "noisy" and far greater sampling frequencies would be required to justify statistical predictions of water quality (47). Also, while bathing sites may be monitored, the places divers go are usually not, and the state of affairs can be quite different at sites which are far from the shoreline. The conditions at these sites may be subject to irregular mixing events which extend to the water-body as a whole; typically the behavioral dependence of point discharges on tide and wind factors as well as on temporal characteristics of the discharges themselves contribute to mixing complexities (48). So, because currently prescribed monitoring procedures are ineffective and the irregularities of pollution dispersion often defeat hydrological modeling at present, the pollution impact at near-shore diving sites cannot be predicted accurately. Even if good monitoring could be done, the unpredictability of dispersive behavior would prevail to frustrate us. For divers then, there is a hazard that they may encounter unpredictable patches of high pollution density at near-shore diving sites.

So the validity of present regulatory practices is being questioned increasingly on the basis of technical aspects and inconsistencies of the underlying epidemiological studies; there is ongoing debate on whether present limits serve as worthwhile microbial water quality standards. A typical technical uncertainty concerns accurate enumeration of organisms made dormant by exposure to stressful marine conditions (49); distortion of the indicator-organism and pathogen relationships will arise if this is not achieved. Another aspect is that international agreement has not been obtained on a "best" indicator organism which can be relied upon to index consistently the health hazards of polluted water (50).

It appears from epidemiological studies despite the many difficulties which afflict them, that there is no appreciable risk of contacting serious illness from polluted waters unless there is gross contamination, and that there is some small risk of contacting minor illness. This accords with the clinical picture given above. Although skepticism has been expressed (51) that significant risk attaches to diving in microbiologically polluted waters, current appreciation of the microbiological hazards to recreational divers warrants a precautionary approach.

Protection at the Individual Level

The view that a diver's equipment offers protection from exposure to pathogens is a mistaken one in the case of recreational divers. The duration of head immersion in polluted water has been shown consistently to mirror rates of illness incidence; it is presumed that the principal infections seen are caused by water entering the ear, nose and throat cavities, and by water ingestion. Divers endure complete water immersion for extended periods and consequently are at enhanced risk from water-borne pathogens. Protective water-tight helmets of the type being worn increasingly by commercial, military and other public agency divers are rarely used in recreational diving. Also, there is risk from exposure to resuspended bottom sediments which can be a substantial pathogen source, and divers should seek to minimize such exposure.

The argument that professional divers are young, fit and healthy, and that their susceptibility to debilitating infection is reduced because of good immunological competence does not extend readily to amateur divers; the health status of the latter category can be very mixed indeed, especially in countries where amateur diving regulations are not enforced or are non-existent. However, divers frequenting home waters will benefit from the immunological competence gained from regular exposure to the prevailing water-borne organisms and through normal contact with members of their local community. The position may be quite different of course, for diving at exotic sites away from home, although frequently such sites are of high amenity value, distant from centers of population and are hence relatively unpolluted.

With regard to personal practices, recreational divers contacting polluted water should seek to reduce water ingestion as far as possible and to disinfect equipment (52). Personal showering and special attention to general hygiene between dives are preventive. Prophylactic ear lavage procedures have been found partially effective in reducing incidences of outer ear infection (19, 53). These measures, though, exact a psychological price from carefree recreational diving.

Protection at the Institutional Level

Divers using recreational diving sites in or close to receiving waters should be informed about the impact of pollution inputs; as outlined above, statutory monitoring regimes are unsatisfactory in this regard. The Irish Underwater Council (Comhairle Fo Thuinn; CFT) with this in mind and reacting to public concern about the health risks to recreational users in Dublin Bay, commissioned a water quality survey of recreational sites in the southern part of the bay (54, 55); the major previous survey (56) did not cover this area. The survey was supported by the Local Authority, CFT and University College Dublin. In summary, the microbiological and aesthetic results for the diving sites indicated that there was cause for concern, though no cause for alarm, concerning water quality. In the absence of diving-related surveys, national amateur diving organizations should consider taking measures to evaluate the microbial hazards of prominent diving sites and to brief local diving officers accordingly. If available, local benthic survey reports should be consulted; the benthic ecology can be a useful indicator of the established pollution status (57, 58).

Responsible bodies have begun to lay down regulations to protect scientific divers, many of whom are of amateur standing; while the 1979 edition of the United Kingdom Underwater Association code of practice for scientific divers (59) was silent on microbiological hazards, the 1987 UNESCO code (60) has sections on sewage contaminated waters, and polluted water and estuarine conditions. Commercial divers in Ireland are circumspect in practice about pollution hazards, but there are currently no relevant national regulations; the 1981 United Kingdom regulations (61) do not cover this aspect either. General enactments in the UK concerning the workplace cover persons "at work"; in Ireland the Factories Act refers only to designated workplaces including docks and quays, so diving locations are not places of work within the Act.

The long-term solution rests with management by the Public Authorities; in Ireland the Water Pollution Act 1977 empowers the local authorities to prepare water management plans for water bodies under their control. The matters discussed here should arise in the course of studies undertaken for development of plans. Recreational diving interests should seek to ensure that such studies encompass the established diving sites, and that improvements made to sewerage installations and action on other pollution sources satisfactorily reduce microbiological hazards for recreational diving.

References

1. Coolbaugh, J.C., O.P. Daily, S.W. Joseph, R.R. Colwell. 1986. Bacterial contamination of divers during training exercises in coastal waters. Mar. Technol. Soc. J. 15:15-21.

2. Shattock, A.G. 1983. Water pollution and hepatitis. pp. 41-45. In: J. Blackwell, F.J. Convery, eds. Promise and Performance: Irish Environmental Policies Analysed. Resource and Environmental Policy Centre, University College, Dublin.

3. Robinton, E.D., E.W. Mood. 1966. A quantitative and qualitative appraisal of microbial pollution of water by swimmers: a preliminary report. J. Hyg. (Camb.) 64:489-499.

4. Shuval, H.I. 1988. The transmission of virus disease by the marine environment. Schriftenr Ver Wasser Boden Lufthyg. 78:7-23.

5. Mosley, J.W. 1975. Epidemiological aspects of microbial standards for bathing beaches. pp. 85-93. In: A.L.H. Gameson, (ed.) Discharge of Sewage From Sea Outfalls. Pergamon Press, Oxford.

6. Moore, B. 1959. Sewage contamination of coastal bathing waters in England and Wales. J. Hyg. (Camb.) 57:435-472.

7. Bryan, J.A., J.D. Lehmann, I.F. Setiady, M.H. Hatch. 1974. An outbreak of hepatitis-A associated with recreational lake water. Am. J. Epidemiol. 99:145-154.

8. Makintubee, S., J. Mallonee, G.R. Istre. 1987. Shigellosis outbreak associated with swimming. Am. J. Pub. Health 77:166-168.

9. Rosenberg, M.L., K.K. Hazlet, J. Schaefer, J.G. Wells, R.C. Pruneda. 1976. Shigellosis from swimming. J. Am. Med. Assoc. 236:1849-1852.

10. Denis, F.A., E. Blanchouin, A. DeLignieres, P. Flamen. 1974. Coxsackie A16 infection from lake water. J. Am. Med. Assoc. 228:1370-1371.

11. Hawley, B.H,. D.P. Morin, M.E. Geraghty, J. Tomkow, C.A. Phillips. 1973. Coxsackievirus B epidemic at a boys' summer camp. J. Am. Med. Assoc. 226:33-36.

12. Baron, R.C., F.D. Murphy, H.B. Greenberg. 1982. Norwalk gastrointestinal illness: an outbreak associated with swimming in a recreational lake and secondary person-to-person transmission. Am. J. Epidemiol. 115:163-172.

13. Philipp, R., E.J. Evans, A.O. Hughes, S.K. Grisdale, R.G. Enticott, A.E. Jephcott. 1985. Health risks of snorkle swimming in untreated water. Internat.. J Epidemiol. 14:624-627.

14. Galbraith, N.S. 1980. Infections associated with swimming pools. Environ. Health 81:31-33.

15. D'Alessio, D.J., T.E. Minor, C.I. Allen, A.A. Tsiatis, D.B. Nelson. 1981. A study of the proportions of swimmers among well controls and children with enterovirus-like illness shedding and not shedding an enterovirus. Am. J. Epidemiol. 113:533-541.

16. Hoadley, A.W., D.E. Knight. 1975. External otitis among swimmers and nonswimmers. Arch. Environ. Health 30:445-448.

17. Reid, T.M.S., I.A. Porter. 1981. An outbreak of otitis externa in competitive swimmers due to *Pseudomonas aeruginosa*. J. Hyg. (Camb.) 86:357-362.

18. Calderon, R., E.W. Mood. 1982. An epidemiological assessment of water quality and swimmer's ear. Arch. Environ. Health. 37:300-305.

19. Alcock, S.R. 1977. Acute otitis externa in divers working in the North Sea: a microbiological survey of seven saturation dives. J. Hyg. (Camb.) 78:395-409.

20. Springer, G.L., E.D. Shapiro. 1985. Fresh water swimming as a risk factor for otitis externa: a case-control study. Arch. Environ. Health 40:202-206.

21. Seyfried, P.L., R.J. Cook. 1984. Otitis externa infections related to *Pseudomonas aeruginosa* levels in five Ontario lakes. Can. J. Pub. Health 5:83-91.

22. Eklu-Natey, D.T., M. Al-Khudri, D. Gauthey. 1985. Epidémiologie de la dermatite des baigneurs et morphologie de Trichobilharzia cf. ocellata dans le lac Leman. Rev Suisse Zoo 92:939-953.

23. DeWailly, E., C. Poirier, F.M. Meyer. 1986. Health hazards associated with windsurfing on polluted waters. Am. J. Pub. Health 76:690-691.

24. Larsen, J.L., A.F. Farid, I. Dalsgaard. 1981. Occurrence of *Vibrio parahaemolyticus* and *Vibrio alginolyticus* in marine and estuarine bathing areas in Danish coast. Zbl. Bakt. Hyg. I. Abt. Orig. B173:338-345.

25. Auerbach, P.S. 1987. Natural microbiologic hazards of the aquatic environment. Clin. Dermatol. 5:52-61.

26. Hanson, P.G., J. Standridge, F. Jarrett, D.G. Maki. 1977. Freshwater wound infection due to *Aeromonas hydrophila*. J. Am. Med. Assoc. 236:1053-1054.

27. Beaune, J., G. Llorca, A. Gonin, Y. Brun, J. Fleurette, J.P. Neidhart. 1978. Phlegmon nécrotique de la main droite, point de départ d'une septicémie due à *Aeromonas hydrophila*. Demonstration de l'origine de l'infection. Nouv. Presse. Med. 7:1206-1207.

28. Fulghum, D.D., W.R. Linton, D. Taplin. 1978. Fatal *Aeromonas hydrophilia* infection of the skin. Southern Med. J. 71:739-741.

29. Joseph, S.W., O.P. Daily, W.S. Hunt, R.J. Seidler, D.A. Allen, R.R. Colwell. 1979. Aeromonas primary wound infection of a diver in polluted waters. J. Clin. Microbiol. 10:46-49.

30. Delbeke, E., M.J. DeMarcq, C. Roubin, B. Baleux. 1987. Contamination aquatique de plaies par *Aeromonas sobria* après bain de rivière. Presse Med. 14:1292.

31. Tonnon, J., J.M. Rives, D. Laniel. 1987. Infection de cuir chêvelu a *Aeromonas hydrophila*, particularités. Presse Med. 16:1244.

32. von Graevenitz, A. 1985. Aeromonas and Plesiomonas. pp. 278-281. In: E.H. Lennette, A. Balows, W.J. Hausler, Jr., H.J. Shadomy, (eds.) Manual of Clinical Microbiology. 4th Edition. American Society of Microbiology, Washington, D.C.

33. Elliot, E.L., R.R. Colwell. 1985. Indicator organisms for estuarine and marine waters. FEMS Microbiol. Rev. 32:61-79.

34. Olivieri, V.P. 1982. Bacterial indicators of pollution. pp. 21-41. In: W.O. Pipes (ed.) Bacterial indicators of pollution. Boca Raton: CRC Press Inc.

35. Brown, J.M., E.A. Campbell, A.D. Rickards, D. Wheeler. 1987. Sewage pollution of bathing waters. Lancet 2:1208-1209.

36. Stevenson, A.H. 1953. Studies of bathing water quality and health. Am. J. Pub. Health 43:529-538.

37. Cabelli, V.J., A.P. Dufour, L.J. McCabe, M.A. Levin. 1982. Swimming-associated gastroenteritis and water quality. Am. J. Epidemiol. 115:606-616.

38. El-Sharkawi, F., M.N.E.R. Hassan. 1982. The relation between the state of pollution in Alexandria swimming beaches and the occurrence of typhoid among bathers. Bull. High Inst. Pub. Health Alexandria 12:337-351.

39. Mujeriego, R., J.M. Bravo, M.T. Feliu. 1983. Recreation in coastal waters; public health implications. pp. 585-594. In: Workshop on pollution of the Mediterranean; VIes journée études pollutions, Cannes. Commission Internationale pour l'Exploration Scientifique de la Mer Mediterranée Monaco.

40. Foulon, G., J. Maurin, N.N. Quoi, G. Martin-Bouyer. 1983. Relationship between the microbiological quality of bathing water and health effects: a preliminary survey. Rev. Francaise Sci. l'Eau 2:127-143.

41. Seyfried, P.L., R.S. Tobin, N.E. Brown, P.F. Ness. 1985. A prospective study of swimming-associated health risk: II-morbidity and microbiological quality of water. Am. J. Pub. Health 75:1071-1075.

42. Fattal, B., E. Peleg-Olevsky, Y. Yoshpe-Purer, H.I. Shuval. 1986. The association between morbidity among bathers and microbial quality of seawater. Wat. Sci. Technol. 18:59-69.

43. Shuval, H.I. 1986. Thalassogenic diseases. UNEP regional seas reports and studies No. 79. Programme Activity Centre for Oceans and Coastal Areas, Nairobi.

44. Moore, B. 1975. The case against microbial standards for bathing beaches. pp. 103-109. In: Discharge of Sewage From Sea Outfalls. A.L.H. Gameson (ed.) Pergamon Press, Oxford.

45. European Economic Community. Council directive 76/160 concerning the quality of bathing water. Off. J. Eur. Comm. L31:1-7.

46. Environmental Protection Agency. 1986. Ambient Water Quality Criteria for Bacteria - 1986. EPA 440/5-84-002. Environmental Protection Agency, Washington, D.C.

47. O'Kane, J.P. 1983. An examination of the EEC bathing water directive. pp. 13-23. In: J. Blackwell, F.J. Convery (eds.) Promise and Performance; Irish Environmental Policies Analysed. Resource and Environmental Policy Centre, University College, Dublin.

48. Masterson, B.F., D. Conry-McDermott. 1989. Use of satellite imagery for the investigation of pollution-related characteristics of Dublin Bay, Ireland. pp. 71-75. In: T-D. Guyenne, G. Calabresi (eds.) Monitoring the Earth's Environment. ESA Publication Division ESTEC, Noordwijk.

49. Roszak, D.B., R.R. Colwell. 1987. Survival strategies of bacteria in the natural environment. Microbiol. Rev. 51:365-397.

50. Dufour, A.P. 1984. Bacterial indicators of recreational water quality. Can. J. Pub. Health 75:49-56.

51. Bachrach, A.J. 1981. Closing remarks. Mar. Technol. Soc. J. 15:57-58.

52. Bond, W.W. 1987. Disinfection of scuba diving equipment. J. Am. Med. Soc. 258:3439-3440.

53. Brook, I., J.C. Coolbaugh. 1984. Changes in the bacterial flora of the external ear canal from the wearing of occlusive equipment. Laryngoscope 94:963-965.

54. Clarke, J., B. Masterson, M. Max, P. O'Connor. 1985. A study of the impact of sewage discharges on the recreational quality of south Dublin Bay. Comhairle fo Thuinn, Dublin.

55. Masterson, B.F., P.E. O'Connor, M.D. Max. 1987. Impact of sewage discharges on recreational diving in Dublin Bay Ireland. Prog. Underwater Sci. 12:213-239.

56. Crisp, D.J. 1976 Survey of environmental conditions in the Liffey estuary and Dublin Bay; summary report. University College of North Wales, Gwynedd.

57. Jeffrey, D.W., J.G. Wilson, C.R. Harris, D.L. Tomlinson. 1985. The application of two simple indices to Irish estuary pollution status. pp. 147-161. In: J.G. Wilson, W. Halcrow (eds.) Estuarine Management and Quality Assessment. Plenum Press, New York.

58. Wilson, J.G. 1987. The Dublin Bay Ecosystem. pp. 21-26. In: M. Brunton, F.J. Convery, A. Johnson (eds.) Managing Dublin Bay. Resource and Environmental Policy Centre, University College, Dublin.

59. National Environment Research Council. N.C. Fleming, D.L. Miles (eds.) Underwater Association Code of Practice for Scientific Diving, Plymouth.

60. Flemming, N.C., M.D. Max (eds.) 1988. Code of practice for scientific diving: principles for the safe practice of scientific diving in different environments. Technical Papers in Marine Sciences 53. UNESCO, Paris.

61. Health and Safety Executive. 1981. A guide to the diving operations at work regulations 1981. Her Majesty's Stationery Office, London.

Prospective Study of Diving-Associated Illnesses

RITA R. COLWELL, ANWAR HUQ, KELLY A. CUNNINGHAM
Department of Microbiology
University of Maryland
College Park, Maryland

GENEVIEVE LOSONSKY
Center of Vaccine Development
University of Maryland at Baltimore
Baltimore, Maryland

Introduction

There are several obvious reasons why the study of diving-associated illnesses is important. *Aeromonas* infection of a diver who sustained a leg wound while diving in the Anacostia River has been recorded in the literature (1). Also, we have evidence from other studies that divers can be rapidly colonized by pathogens from the environment (2). In addition, Chesapeake Bay watermen have exhibited higher antibody titers to *Vibrio vulnificus* compared to people not routinely exposed to the Bay (3). Consequently, we are studying whether organisms that colonize divers are persistent or whether they are just transient colonizers that do not cause problems.

Methodology

For such a study, volunteers are required; a variety of diving teams has been enlisted, including Navy divers working in Panama City, Florida, at the Naval Diving Training and Salvage Center. A very active group of volunteers has been recruited from the Washington, D.C., Metropolitan Police Department, Metropolitan Washington National Airport Authority, and Underwater Recovery Unit of the Baltimore County Police. Our proposal and protocols had to be approved not only by human volunteer research committees at the University of Maryland, but also by the Navy and the respective Police Departments. Both of those processes took much longer than anticipated because of the need for modifying the protocol and gaining approval for each modification.

Part of our study has involved setting up a questionnaire that volunteers can answer, enabling us to evaluate their health status up to the study period and throughout the study.

We give the questionnaire to the volunteers before they begin their dive training and also approximately one month after they have been diving so that their health status can be assessed throughout the study. The questionnaire inquires about wound, ear, nose, throat, gastrointestinal infections, respiratory infections, headaches, and blurred vision. Volunteers simply circle a yes or a no if any of these symptoms or ailments apply to them and then describe how long these symptoms persisted and whether or not they required treatment. Another question covers accidents or injuries that have occurred to the divers, to detail their health status when joining the study and during the sampling period. In addition to questions that appear on the protocol during the physical exam, the physician also asks additional questions -- some data that are useful to have, such as if the diver is a smoker, length and duration of smoking, etc. We also ask about previous diving experience, because quite a few of the divers in these training courses are new divers; that is not the case for all of them.

Some of them have been recreational divers, and, in one case, a Navy Seal diver with extensive experience, and, therefore, potential prior exposure to polluted water was included in a study set. We also ask the volunteers whether or not they irrigate their ears with solution to prevent ear infections. Although this is a well-known remedy, it remains a personal option, and we need to know what effect this might have on the ear swab samples that we take.

As well as microbiological hazards (our primary focus), the study also analyzes chemical hazards, which also influence the nature of the protocol, though the emphasis is on exposure to pathogenic microorganisms. One final question we asked the volunteers is about alcohol consumption in the past 24-72 hours. One of the tests we do, a blood test for liver enzyme functions, can be affected by alcohol.

At the beginning of the study, an attempt was made to isolate the potentially pathogenic microorganisms from water, including coliforms and pathogenic *E. coli*, in particular, opportunistic pathogens such as *Serratia* and *Proteus* spp. Both Dr. Colwell and Dr. Joseph, in this session, mentioned problems of culturing anerobes and their potential ability to cause gastroenteritis. There is also an oxidase positive group of gram negative bacteria, most notably *Aeromonas hydrophila* in particular, and several species of *Vibrio* which pose problems for human health. In addition to the gram-negative bacteria, there are several genera of gram-positives that can potentially cause wound infections, as well as a host of viruses and protozoans that are implicated in gastroenteritis and meningoencephalitis. Thus, there are numerous problems that could potentially be investigated and, obviously, one study cannot address them all at the same time. The bacteria selected for study in this project include: *Aeromonas* spp., *Aeromonas hydrophila* in particular, because of previous history of infections and illnesses that have been caused by this pathogen; *Vibrio cholerae* and *Vibrio parahaemolyticus*, two examples of organisms autochthonous to estuaries that can potentially cause infections; and *Pseudomonas aeruginosa*, which is the primary agent in swimmer's ear or ear infections, in general.

To explain the protocol used, the procedure followed on-site with the divers is described as follows.

At the beginning of the study, after the volunteers have agreed to be part of the study, they fill out a consent form and questionnaire. The questionnaire is again completed later in the study, as well. The volunteers also supply several clinical specimens such as stool and urine samples. We use different enrichment procedures for bacteriological analysis of stool samples and have targeted three of our five organisms of interest, namely, the three that can cause gastroenteritis: *Vibrio cholerae*; *Aeromonas hydrophila*; and *Salmonella*. We use four different selective plates, including TCBS, which is a medium selective for *Vibrio*. From that medium, yellow colonies, which indicate sucrose fermentation, signal the presence of *Vibrio cholerae*. Yellow colonies from TCBS plate are presumptive *Vibrio cholerae*, confirmed by additional analyses.

Aeromonas hydrophila are relatively easily isolated and produce bright purple colonies on a selective medium. A selenite-16 broth XLD medium is used, and color reactions on the medium indicate potential *Salmonella* isolates. Black centered colonies on XLD and lactose negative, whitish colonies are considered presumptive *Salmonella typhimurium*. All the isolates are picked and placed in a storage medium for future study.

The study of the divers includes a physical examination. A medical doctor carries out the examination and looks primarily for ear, nose and throat problems, tissue redness, irritation, conjunctivitis and laryngitis. The doctor also conducts a pulmonary examination, which is carried out before and after the dive. We collect blood samples to measure antibody titer over the course of the dive training, because, as has been mentioned earlier, one of our primary goals is to determine whether or not the divers develop immune response to any of the pathogens or potential pathogens colonizing them and also to evaluate kidney functions and blood count using liver enzyme function test. The last three tests, blood count, liver function and urine specimen, are portions of the protocol designed to look at the effects of acute chemical toxicity, as opposed to microbiological hazards. The kidney, the liver, and the blood components are the most obviously affected areas when analyzing potential exposure to hazardous chemicals.

We swab the divers' ears, nostrils, throat, as well as gear and mouthpiece with sterile cotton swabs, placing the swabs into special transport media. At the laboratory, we again streak a series of selective plates. On specific agar media, blue/green colonies indicate the production of a blue/green pigment which signals probable *Pseudomonas* isolates. In addition to those already discussed, we pick other isolates from *Pseudomonas* agar plates and TCBS agar to screen for *Vibrio vulnificus*. There are sucrose nonfermenters which show up as green colonies on TCBS plates, and we identify those isolates also.

Water and sediment samples are collected on the day of the dive at the time the volunteers are diving. This sampling takes place while the divers are experiencing the given environmental conditions, to measure the actual bacterial load at the dive site and do quantitative analyses, including acridine orange direct counts (AODC), direct viable counts (DVC), and standard plate counts, again, to determine the number of organisms per volume of water present at the dive site.

In addition, there is another procedure, which involves enrichment and selection for potential isolates of interest, i.e., inoculation of an *Aeromonas* enrichment broth for *Aeromonas hydrophila* and alkaline peptone water for *V. cholerae* and other vibrios, as well as a selenite-16 broth for the *Salmonella*.

One more point of protocol is the time schedule. The timeframe is a bit different for the different samples. After the swabs are collected, we also perform a physical exam before the dive and immediately after the dive or at least within two hours. The urine, stool and blood samples for liver function test are collected about 24 to 48 hours after the dive and analyzed afterwards to give a response time, since exposure to an acute chemical does not show up immediately. Approximately 24 to 48 hours after exposure is the time within which changes in liver enzymes or kidney functions would be observed. The stool specimen is collected approximately 24 to 48 hours after the dive because ingested pathogens must traverse the intestinal tract to exhibit evidence of infection.

We collect blood for serology right before the dive and one-month later, to monitor antibody titers. Also, if the volunteers are exposed to an organism repeatedly during dive training, this would essentially boost the titer (if it is going to increase), and the titer will increase significantly over a one-month response time. As mentioned above, water and sediment samples are collected on the day and at the time of the dive.

We have completed several studies to date. In February of 1987, we conducted our first study in Norfolk, Virginia, as a part of a research cruise where we collected water and sediment samples and isolated organisms, following the protocols described above. We collected approximately 120 isolates and completed a preliminary analysis of water and sediment.

Following that study, in July of 1987, we travelled to Honolulu, Hawaii, and worked with NOAA divers. This represented our first run through the protocol using a small group of three divers. We checked out the protocol and collected baseline data. The water where the volunteers were diving was very clean. The number of curable organisms per ml was low, on the order of $10^3 - 10^4$ per ml. Preliminary biochemical testing showed that most isolates represented the normal flora of the skin, primarily gram-positive cocci. We did not isolate pathogens from the divers. Thus, the study in Hawaii, although based on a small group of divers, served its purpose and provided good baseline data.

A study was conducted in Panama City, Florida, at the Naval Diving Training and Salvage Center, in February, 1988, with a medical doctor and an aquatic toxicologist on the team. Basically, an operation to begin experimental sampling with the volunteers was set up.

In June, 1988, at the Diving Training Salvage Center in Panama City, we initiated our first formal study with Navy volunteers. We collected a total of 700 isolates from 10 divers and conducted water sampling. The sampling went well, the protocols worked, we had excellent cooperation at the dive center, and the microbiology laboratory (through the assistance of Dr. J. Spain) and the hospital at Panama City in processing blood and urine specimens. The divers were enthusiastic about the study and saw benefits in knowing what hazards, if any, they encountered.

If we isolate an organism from a diver, i.e., if we isolate one of the five species targeted for study, we determine if the organism is also isolated from the water. If it was isolated from the water, there is at least preliminary evidence of an association, that what the divers encountered in the water was what they were colonized with. The important question is whether or not they developed an immune response. At this point, we analyze the serum samples (which we keep in storage) and determine the diver's antibody titer before dive training and one month after dive training, to see if there is a difference and if it has changed over time.

Conclusion

Results show that the majority of experienced divers (71-86%) studied have pre-existing antibody to the potential waterborne pathogens they are exposed to. Divers can become infected with waterborne organisms as evidenced by seroconversion water isolates at the dive site, but this is only a limited range of water organisms the diver can be ultimately infected with.

In summary, this project is designed to determine microbiological hazards encountered by divers in polluted waters and should prove to be a useful paradigm for future studies.

Acknowledgment

This study was funded by National Oceanic and Atmospheric Administration Grant No. NA86AA-D-SG006, 10.

References

1. Joseph, S.W., O.P. Daily, W.S. Hunt, R.J. Seidler, D.A. Allen, R.R. Colwell. 1979. *Aeromonas* primary wound infection of a diver in polluted waters. J. Clin. Microbiol. 10:46-49.

2. Seidler, R.J., D.A. Allen, H. Lockman, et. al. 1980. Isolation, enumeration and characterization of *Aeromonas* from polluted waters encountered in diving operations. Appl. Environ. Microbiol. 39:1010-1018.

3. Lefkowitz, A., G.S. Fout, G. Losonsky, S.S. Wasserman, E. Israel, J.G. Morris. In press. A serosurvey of pathogens associated with shellfish: Prevalence of antibodies to *Vibrio* species and Norwalk virus in Chesapeake Bay Region. Am. J. Epidemiol.

Section 2

General Contamination

Evolution of the National Oceanic and Atmospheric Administration's Capabilities for Polluted Water Diving

WILLIAM C. PHOEL
National Oceanic and Atmospheric Administration
National Marine Fisheries Service, Sandy Hook Laboratory
Highlands, New Jersey

J. MORGAN WELLS
National Oceanic and Atmospheric Administration
NOAA Diving Program
Rockville, Maryland

Introduction

The National Oceanic and Atmospheric Administration (NOAA) became interested in diving in polluted waters in the early 1970s when, as a result of heightened awareness of the problems of ocean dumping, the National Marine Fisheries Service laboratory at Sandy Hook, NJ, began to conduct *in situ* investigations at several disposal sites in the New York Bight. Shortly thereafter, other elements of NOAA (e.g., National Ocean Service and Atlantic Oceanographic and Meteorological Laboratory) also began to work in these contaminated areas. In the mid-1970s the NOAA Diving Program began tests and evaluations of several commercially-available suit and helmet combinations for protecting divers from biological, chemical, and radiological water-borne hazards. Other federal agencies (e.g., Environmental Protection Agency, U.S. Navy, and U.S. Coast Guard), as well as universities, medical groups, and private diving companies also were concerned and contributed to meetings, workshops, and demonstrations of the equipment and procedures as they were developed.

Polluted Water Diving

Before 1970, polluted water diving was not generally considered to be a problem for researchers. Scuba diving was the modus operandi for underwater scientists, but almost all research diving was accomplished in relatively clean waters where animals, plants, and geological features of interest could be observed, photographed, and manipulated. Exceptions occurred when research-oriented scuba divers installed sampling or measuring instrumentation (e.g., tide and wave height gauges) in harbors and ports that had water of dubious quality. Commercial and military divers who routinely work in waters of poor quality were usually well protected by helmets and dry diving dress. For waters polluted by radiation or certain chemicals, however, even the commercial equipment and procedures was inadequate. Quantitative data, or even anecdotal information, on the incidence of sickness or injury to research divers in water that is polluted biologically, chemically, or radioactively were not available until recently, so the extent of problems caused by diving in polluted water using scuba before 1970 cannot be determined.

Between 1970 and 1975 public and scientific interest in man's impact upon the rivers, estuaries, coastal waters, and open ocean increased dramatically. Congressional legislation and federal agencies mandated that researchers include these heretofore avoided environments in their scientific investigations. Quoting from the Northeast Monitoring Program (1):

The National Oceanic and Atmospheric Administration (NOAA), as the nation's principal civilian ocean agency, has a commitment to determine the effects of man's activities on coastal/estuarine waters, the ecosystems contained therein, and their resources. Part of this commitment is to develop a database, through long-term monitoring, that will allow the assessment of the effects of pollutants on ecosystems and resources, and will enable early detection of and response to significant environmental changes.

This policy statement, coupled with NOAA's capabilities and research responsibilities for investigating the more acute and often dramatic introductions of pollutants into water, makes it evident that research, implicitly including *in situ* research, is required.

Limited visibility in areas such as dumpsites, petroleum or toxic substance spills, and turbid harbor and coastal waters often precludes the underwater scientist from accomplishing the usual tasks of identifying, counting, and photographing the living and nonliving items of interest, but many other tasks can be accomplished. For example, underwater instrumentation for measuring current speed and direction, sediment oxygen demand and nutrient regeneration, and other *in situ* experimentation can be set up without the aid of sight. When visibility was zero, the authors have used the sense of touch to (a) retrieve specimens during a fish kill, (b) determine the relative abundances of live and dead oysters and clams, and (c) collect sediment samples from precise locations. Feeling about in soft mud with zero visibility and many submerged obstacles is unpleasant and requires caution but is often necessary. The physical, biological, chemical, and perhaps even radiological environments of these areas dictated that underwater scientific investigation be conducted using equipment and procedures more characteristic of commercial and military diving, rather than scuba.

MESA Program

With the requirement and responsibility for conducting research into man's effect on the nation's waters firmly established, NOAA responded with the Marine Ecosystems Analysis Program (MESA). In the northeast region of the United States, the MESA Program established the New York Bight Project and, beginning in 1973, depended on expertise at the National Marine Fisheries Service Laboratory at Sandy Hook, NJ. The New York Bight not only received large inputs of biological and chemical material through nonpoint sources but also contained a dredge spoil site, industrial acid waste site, and the nation's only offshore disposal site for municipal sewage sludge. (By January 1988, the New York Bight sewage sludge dumpsite was closed and the sludge was being dumped off the continental shelf approximately 110 miles east of Cape May, NJ. In January 1989, the last company at the acid waste dumpsite announced that they would cease dumping shortly. The research reports and publications generated by the various environmental programs undoubtedly contributed to the decreased anthropogenic input to the New York Bight.)

Hazardous Diving Gear

In 1973, NOAA had no experience in polluted-water diving and no coordinated policy for diving in polluted water; it was up to the diving officer of each individual laboratory or unit to address the problem. The first system used by NOAA specifically for diving in polluted waters was developed by W. C. Phoel while he was Diving Officer at the Sandy Hook Laboratory and was based on his military and commercial diving experience (2). The system was developed to protect the divers while accomplishing *in situ* research in the New York Bight dumpsites. While basically concerned with microbial hazards, it also proved effective against some chemical contamination.

The criteria for the mask or helmet were that they provide complete head isolation from polluted waters, be capable of water-tight connection to a suit, be capable of both demand and free-flow modes, be usable with scuba or surface supply, provide good communications, be commercially available or "off the shelf," and be affordable. Criteria for the suit or diving dress were that they were to be completely dry, preferably had variable volume, were capable of mating dry to the mask or helmet, allowed the diver to swim, could be washed and sanitized easily, were commercially available or "off the shelf," and were affordable.

After 1975 NOAA divers were actively working in polluted areas, so the NOAA Diving Program initiated studies into the effectiveness of the equipment and associated procedures for protecting divers from pathogenic microorganisms. These studies, which confirmed that microbial pathogens posed potential hazards for the divers, were conducted in collaboration with the Naval Medical Research Institute and the University of Maryland (3,4). As interest in the problem of diving in polluted water spread among universities, commercial diving companies, and government agencies such as EPA, U. S. Navy, U. S. Coast Guard, and Department of Energy, it became apparent that the hazards of diving in water which is polluted with chemicals and radioactivity would also have to be addressed. In 1982, an informal group of representatives from government, academia, and industry convened a workshop to address such hazards (5).

Initial work concentrated on the structure and function of diving equipment to isolate the diver and the breathing gas from contaminated water. Modifications to commercially available equipment were required. That these undertakings were important and timely was demonstrated in July 1982 when 17 of 40 New York City firemen contracted an enteric parasitic infection that causes diarrhea and occasionally more serious symptoms while practicing scuba diving in the Hudson River near Battery Park and the World Trade Center (6). Subsequent investigations by the New York City Fire Department and New York City Department of Health determined that there was a causal relationship between scuba diving in water contaminated by sewage, and enteric protozoan infection and gastrointestinal illness (7).

Gear Testing

In 1982, an interagency agreement between NOAA and EPA provided financial support for NOAA to evaluate commercial and modified diving equipment for use in contaminated water. The first series of tests were conducted between April 1982 and March 1983 at the U. S. Naval Surface Weapons Center, White Oak, MD, and involved various combinations of helmets and suits. The testing facility consisted of a 50-foot diameter, 100-foot deep water tower with a grated platform that could be raised or lowered to precise depths. Because the equipment combinations were experimental and the tests included blowups or uncontrolled ascents, there was a possibility of gas embolism. For safety, a NOAA 42-inch diameter recompression chamber was maintained on site.

Results of these tests, coupled with the earlier work, indicated that a tiered approach to polluted-water diving was practical and still afforded the diver protection. The first tier was developed for use in waters which are minimally polluted with pathogens. At this time no quantitative limits can be used to describe "minimally" or "heavily" contaminated water; however, quick-response kits are being developed to measure pollutant levels. The kind of chemicals or other contaminants involved and the maximum concentration of each in the water are the prime considerations when deciding which level of protective system will be used. EPA has published the maximum exposure levels of many contaminants and also a list of toxic chemicals which precludes the use of divers. In minimally-contaminated waters where the use of an umbilical was precluded (e.g., fighting fires under piers), a full face mask that maintained positive pressure within the mask, coupled with a dry suit with a smooth exterior of vulcanized rubber, was considered adequate. The configuration of this type of mask prevents inhaled air from passing over the water droplets normally present in the exhaust port. Scuba regulators not having this feature allow inhaled air to pass over the contaminated water in the exhaust port, creating a mist that can be inhaled and/or ingested by the diver. It was also determined that suits with rough exteriors, such as nylon or neoprene, were extremely difficult or impossible to disinfect after use in polluted water.

The second tier was for waters contaminated with pathogens and some chemical pollutants (e.g., oil or gasoline). The system provides surface-supplied air to a helmet capable of either demand or free-flow modes and which completely isolates a diver's head from the water. The helmet is mechanically attached to a vulcanized rubber suit to minimize potential leaks. The Navy's MK 12 diving system provided the basis for the system to be used in "heavily" contaminated water, the third tier.

The systems developed at White Oak were tested in March 1983 at the EPA facility in Leonardo, NJ. Divers exercised and worked in a 5000-gallon test tank to which fluorescent dye and ammonia had been added to simulate a chemical spill. During the dives, one-piece cotton "bunny" suits were worn beneath the diving suits, cotton caps were worn under the helmets, and cotton swabs were carried inside the helmets. After the dive was completed the bunny suit and cotton cap were examined with ultraviolet light to determine if any leaks had occurred. The cotton swabs in the helmet were analyzed to determine if ammonia had penetrated into the air the divers were breathing. During these tests it became apparent that equipment and procedures were needed to protect surface personnel, especially those tending the divers, from contamination. Also, equipment and procedures to decontaminate the divers, tenders, and other surface support personnel had to be developed.

Gear Innovations

During 1983, the NOAA Diving Program, under the direction of J.M. Wells, analyzed data collected during the tests, and redesigned, fabricated, and bench-tested the equipment. Protocols and future experiments were developed and designed. Two of the more innovative pieces of equipment developed during this time were the "series exhaust valve" (SEV) and the "suit under suit" (SUS). The SEV, an arrangement in which two exhaust valves are aligned in series, solved the problem of "splashback" through the exhaust valve of demand regulators. Splashback of contaminated water into the diver's helmet should not occur, so the use of otherwise acceptable commercially available helmets was precluded until the SEV was developed. The knowledge that many divers working in contaminated environments would also be working in very limited visibility around sharp, jagged metal from shipwrecks, tanker trucks, railroad tank cars, and the like, raised concern about tearing or puncturing the protective diving dress. To alleviate this concern the SUS was designed and fabricated; it consists of a tight fitting, one-inch thick, black neoprene suit with attached booties under a modified vulcanized rubber dry suit. The modification permits the suits to be joined at the neck by a clamp to create a closed space between the two suits. Further modifications permit clean water (usually fresh), supplied via umbilical from a surface source (e.g., ship, barge, or dock), to enter the outer suit through a valve on the chest in order to flood the space between the suits. One-way outlet valves on the left arm and at both ankles permit the constant supply of water to exit. A garment of heavy cloth is worn over the vulcanized rubber suit for protection against chafing and to restrain the suit from blow-up. Should the dry suit tear or be pierced, the positive pressure of injected clean water between the suits will prevent contaminated water from entering and reaching the diver.

The next series of tests and evaluations were conducted at NOAA's Experimental Diving Unit in December, 1983, and

February, 1984. The facility was then at the National Marine Fisheries Service Miami Laboratory on Virginia Key and included a diving tower (12-foot high by 12-foot diameter) in which the water temperature can be regulated. The divers participating in the tests wore electrocardiogram leads and temperature sensors so that their heart rate and body core temperature could be monitored during exercise.

Divers working in warm water and totally encapsulated from the contaminated environment exhibited signs of heat stress early in a dive, with the onset of signs and symptoms dependent on the work load and ambient water temperature. Although the diver perspires profusely, no cooling is possible since evaporation is prevented by the suit. Further testing conducted in water heated to 44°C indicated that a diver wearing the SUS could perform the same work for at least 3 times as long and not show the effects of heat stress exhibited by the diver wearing conventional protective equipment (8, 9). Conversely, if a diver is working in cold water, warm water can be supplied into the suit.

Research on the protection and decontamination of surface support personnel continued at the NOAA facility and resulted in procedures and equipment capable of addressing various levels of contamination. Two systems for diving in contaminated water evolved. One, for use in water polluted with pathogens and certain chemicals (e.g., oil), consisted of a modified demand/free-flow helmet mated to a modified vulcanized rubber dry suit. Both helmet and suit are commercially available but are modified by the NOAA Diving Program for use in contaminated water The other system is designed for use in water contaminated by chemicals and limited radioactivity and consists of the SUS adapted to a modified Navy MK 12 diving helmet and dress.

In September, 1984, at NOAA's Western Regional Center (Seattle, Washington), NOAA personnel experienced in polluted water diving were provided an opportunity to use and evaluate their equipment. The week before the dress rehearsal was to begin, the divers and surface support personnel assembled the diving systems, set up decontamination stations, established communication and underwater video headquarters, and partially buried 55-gallon drums half-filled with sand in the mud and sand bottom of Lake Washington. The drums, used to simulate a chemical spill, would be searched for, dug out, placed in special oversized containers, and then lifted from the lakebed.

The dive system for waters polluted by pathogens was tested by using an underwater torch to cut metal, install a tide gauge, and set up underwater incubators for scientific experiments. The SUS system was tested by searching for, packing and retrieving the buried 55-gallon drums and using an underwater suction hose to "vacuum" a simulated spill of polychlorinated biphenyl.

Problems of compatibility and permeability of suit, helmet, and hose materials by some substances remain to be solved. Some chemicals, especially in high concentrations, may be capable of dissolving protective suits. Other chemicals, for example bromine, methyl parathion, and chlordane, present too much risk to divers in any protective system.

Remotely operated vehicles (ROVs) have shown great promise in the reconnaissance of a spill site, but some chemicals, depending on concentration, can be harmful to ROV components such as umbilical and seals. ROVs also can be difficult to maneuver around obstacles; should an ROV become entangled, a diver would be required to retrieve it. Unless the ROV is considered expendable (an expensive consideration), its use is limited to waters in which a diver can safely operate. Of course there may be spills in which the cost of an ROV is minimal compared to the total cost of the operation.

Operationally, the newly-designed diving systems have been used to search for drums suspected of leaking toxic wastes at the abandoned Big Gorilla open pit coal quarry near McAdoo, PA, and to support scientific investigations in the Palau Islands in a lake which had extremely high levels of hydrogen sulfide dissolved in the bottom water. Fulfilling one of NOAA's responsibilities for technology transfer, several commercial diving equipment manufacturers are using the innovations developed by the NOAA Diving Program, with support from the EPA and other government agencies, and are now offering off-the-shelf equipment for diving in contaminated waters.

References

1. Northeast Monitoring Program (NEMP). 1981. Annual NEMP report on the health of the northeast coastal waters of the United States, 1980. NOAA Tech. Mem. NMFS-F/NEC-10.

2. Phoel, W.C. 1978. A diving system for polluted waters. pp. 232-237. In: P.S. Riegel (ed.) Proceedings of the Working Diver 1978 Symposium. Marine Technology Society, Columbus.

3. Colwell, R.R. editor. 1981. Special issue on microbial hazards of divers in polluted waters. Marine Technology Society Journal 15(2).

4. Colwell, R.R. editor. 1982. Microbial hazards of diving in polluted waters. A proceedings. College Park, MD: Sea Grant Program, University of Maryland. Publication No. UM-SG-TS-82-01.

5. Lotz, W.E. editor. 1983. Protection of divers in water containing hazardous chemicals, pathogenic organisms and radioactive material. Undersea Medical Society, Bethesda, MD.

6. Goodman, A., S. Schultz, E. Bell, et al. 1983. Gastrointestinal illness among SCUBA divers - New York City. MMWR, Centers for Disease Control 32:576-577.

7. Jones, C.J., A. Goodman, S. Friedman, S. Schultz. 1985. SCUBA diving in polluted coastal waters. pp. 959-961. In: Record of Ocean Engineering and the Environment Conference. Marine Technology Society, San Diego.

8. Stringer, J. 1985. Surviving underwater deathtraps. U.S. Department of Commerce, NOAA, 15:1.

9. Wells, J.M. 1984. Equipment innovations cut risks for divers in polluted waters. Sea Technology 25(12):22-23.

Intervention in Hostile Environments: Comex Experience

J. P. Imbert, Y. Chardard, and P. Poupel
Comex S. A.
36 Boulevard des Oceans
13275 Marseille, Cedex 9, France

Introduction

Most professional divers start their careers as sport divers, dreaming of the clear water and the golden sand of tropical islands. In the offshore industry, inevitably, they find themselves in the cold and dark waters of the North Sea, working on steel structures with nothing but a remotely operated vehicle for company. The worst jobs come from inland or civil engineering diving. Divers have worked in the lower part of the huge refinery gas tanks where the water gets up to pH 4 due to gas dilution. They have inspected pools of nuclear plants and have been exposed to radiation. Most of the time, they drag their umbilical in the murky water of a sewage gallery or a drain canal. These divers are very proud of their jobs because intervention in hostile environments requires special equipment and techniques in addition to highly professional experience. This paper presents some examples of operations that have built Comex experience in hostile environments.

Comex Pro PolluSuit

From the old hard-hat diving system, diving suits have evolved to modern dry suit and helmet systems which have been adapted successfully to intervention in polluted water. The basic safety principle of all intervention is to isolate the diver from the environment. The Comex Pro Pollusuit is a dry suit that is perfectly watertight because the boots and gloves are welded to the suit. The diver dons the suit by way of a waterproof zipper placed on the back. However, the originality of the system lies in the diving helmet, which was derived from a deep diving system that uses a gas-reclaim breathing unit and heat shroud for gas heating.

The gas-reclaim breathing unit allows the dumping of exhaled gas to the surface through a return line and thus avoids the use of any mushroom valve for exhaust gases, as in traditional helmets, which are never 100% tight and are the main cause of water ingress and pollution inside a diving helmet. The breathing unit has a membrane to trigger the gas delivery on demand, and defective membranes are known to be the second source of water leakage on a helmet. To overcome this risk, the breathing unit was fitted with a heat shroud, which initially was a system developed for gas heating during deep diving operations. The shroud is simply flushed with fresh water and provides the breathing unit with an envelope of nonpolluted water to prevent accidents in case of membrane leakage.

Intervention in Hydrocarbon-Polluted Waters

Hydrocarbon pollution is common during commercial diving operations. Divers often have to work on damaged risers in a mixture of crude oil and drilling mud. They routinely change flexible hoses on SBM terminals after they have been damaged by a tanker and crude oil has spilled. Sometimes, a wrecked tanker must be salvaged. Hydrocarbon pollution hazards depend on the crude oil composition but generally include: (a) skin problems when in contact; (b) damage to the rubber or neoprene parts of the breathing equipment, such as the exhaust mushroom valve, the regulator, or the neck dam (these parts get swollen and become more fragile or no longer watertight; the consequence may be equipment malfunction or water ingress; intoxication may occur if hydrocarbons are ingested along with sea water); and (c) degassing and production of toxic vapors due to the difference in temperature when introduced into a chamber.

The vapors vary from one deposit to the other but mainly involve CO_2, hydrogen sulfide, mercaptans, aromatic hydrocarbons, and saturated aliphatic and alicyclic hydrocarbons. Fortunately, these chemicals are adequately absorbed by sodalime and activated charcoal which can be easily included in the filters of the bell and chamber scrubbers.

For shallow operations based on surface-supplied diving, divers generally use standard dry suits fitted with a diving helmet. As a precaution, the standard diaphragms and valves are replaced with silicone ones that are resistant to hydrocarbons, and the equipment is rinsed and checked after each dive. For deeper operations, bell diving is required, and specific procedures must be defined to provide divers with reasonable working conditions, attempt to avoid pollution of the bell and the chambers, and control the pollutant in the bell and chambers by adapting the scrubber systems.

In 1976 the *Boehlen*, a supertanker broken and sunk by a storm off the Brittany shore, caused a major ecological disaster on the western coast of France. The wreck rested on the seabed at 90 meters; its tanks were hemorrhaging crude oil to the surface. The French government contracted Comex to pump out the remaining oil and stop the black tide. At such a depth, work could only be carried out by saturation diving, and involved 6 months of difficult operations. Divers had to access the tanks to install hot water flow lines and suction hoses. The crude oil grade was of "Boscan" type, which has the consistency of warm butter and is very sticky. It was inevitable that the divers became entirely covered with a layer of crude oil several centimeters thick. The procedures were first aimed at reducing the amount of crude oil entering the bell. Divers were wearing disposable overalls over their diving equipment which they took off at the end of the dive, before reentering the bell. Equipment such as suits, fins, gloves and tools were stored outside the bell. Once in the bell, the helmets and the umbilicals were sealed in large plastic bags with silver tape to minimize degassing. The bell scrubber was then refilled with charcoal and the diver and the bellman used emergency rebreathers which utilized cartridges filled with charcoal.

When the bell was recovered at surface, a two-part rubber diaphragm was installed in the trunking between bell and chamber. During the transfer under pressure, a slight over-pressure was created in the chamber to produce a gas flow from the chamber to the bell preventing any atmosphere mixing. The diver would then transfer through the diaphragms rapidly. Inside the chamber, the environmental control unit filters were also filled with equal parts of sodalime and activated charcoal to absorb any remaining chemical. After the transfer of the divers, the bell was decompressed to surface. The bell and the diving equipment were thoroughly cleaned and rinsed with aliphatic-saturated solvents. The cleaning agent used was Gamlen Chemical Co ref. OSR LT 126, which is entirely emulsifiable and is approved by the British Department of Industry. The overall procedure proved to be efficient; the chamber atmosphere was monitored by gas chromatography, and no problem of chamber pollution was recorded.

Intervention in High Temperature Environments

Difficult diving operations like the *Boehlen* salvage have made diving contractors develop an expertise, now well recognized by the nuclear industry, for interventions in the pressurized cells of nuclear plants during qualification tests or diving in a reactor pool for maintenance and debris removal. One of the recent Comex developments has been the Hades suit program supported by the French National Electricity for manned intervention in high temperature environment.

The development was related to the Super Phoenix project at Creys Malville nuclear installations. A unique example of a commercial fast-breeder nuclear reactor, the reactor main shell is doubled with a safety tank, both shells being made of high quality stainless steel. The interspace is made inert with pure nitrogen to avoid corrosion risk. Super Phoenix was stopped in 1985 for technical reasons and during inspection work, a small robot operating in the inner space jammed and left debris of graphite between the two shells. An attempt to clear the debris proved unsuccessful and human intervention was required in the narrow space of the intertank gap.

Intervention time was set to one hour and the specifications for the job were as follows: (a) Access to the intertank space was through a 3-m concrete floor in a 450-mm diameter trunking. The distance in between the two shells was only 700 mm. (b) No radioprotection equipment was required because the reactor had not been restarted since 1985 and the remaining radioactivity was low. (c) The inner tank, containing the fuel, was filled with liquid sodium, maintained at a constant temperature of $180°C$; temperature in the inner space ranged from $80°C$ to $150°C$. (d) The intertank was made inert with nitrogen which could not be polluted by any gas leakage.

The program aimed at developing an active thermal protection suit called Hades, which permitted man to work in the inner space of the Super Phoenix double shell. A first prototype was developed in 1986, consisting of (a) a suit with closed circuit air ventilation using high-strength Kevlar fabric with special coating to avoid introduction of any oxygen or halogen pollution of the inert space; (b) an air free-flow breathing system with built-in over-pressure and gas reclaim; (c) a safety harness integrated to the suit to allow the diver to be lowered using a small winch; and (d) an umbilical supplying the diver from the surface with breathing gas, cooling gas, and communications. The Hades system was first demonstrated in a full-scale mock-up of the reactor at ambient temperature to test the ergonomic performance of the suit. The diver was lowered down to 35 m and was able to work and reach the bottom of the hull.

Trials at increased temperatures were carried out in an oven to test the actual thermal protection provided by the suit. It was discovered that the umbilical had too large a heat-exchange surface so that it warmed the gas on the way to the diver. Several modifications had to be introduced following these trials, and the suit was redesigned in 1988: (a) The final version of Hades suit had a double system of active heat protection. The inner air-ventilated suit was placed within an overall suit consisting of a bi-layer of Kevlar and Nomex garments. The bi-layer was flushed with nitrogen in open circuit and proved to be very effective in protecting the diver. (b) The umbilical was better insulated. (c) An emergency bail-out system was added to the suit providing a breathing autonomy of 10 min, sufficient for the diver to be recovered at surface. (d) A complete monitoring system (oxygen level, 3 skin thermal sensors and suit over-pressure) was installed to monitor the thermal state of the diver and control the operation of the suit. (e) A special video camera, featuring a micro CCD camera and a nitrogen cooling system, was developed to control the diver and the job. After subsequent trials, the Hades suit was finally used in October 1988. The diver carried out the inspection of the upper part of the Super Phoenix tank at a temperature of $80°C$ with full success.

Conclusion

Although the experience reported is only part of the Comex expertise, it can be seen that commercial diving technology does not apply only to subsea operations and can be successfully implemented in other industries. Intervention in hostile environments certainly started with cold, dark, and stinking waters but has led the diving industry to win new markets in the chemical industry, in nuclear plants, and in the space industry, which, at least in Europe, is the next challenge.

Hyperthermia in Divers and Diver Support Personnel

J. MORGAN WELLS
NOAA Diving Program
National Oceanic and Atmospheric Administration
Washington, D.C.

Introduction

The encapsulation required to protect divers and surface personnel from harmful pathogenic microorganisms in tropical, temperate, and artificially-heated contaminated environments can result in hyperthermia. To compound the problem, decontamination procedures often extend the length of time divers remain suited, often while they stand in the hot sun, and may limit the aid that can be given to them if they are hyperthermic. Tenders equipped with protective suits and breathing apparatus are also prime targets for hyperthermia.

The NOAA Dive Program conducted a series of experiments which showed that certain types of equipment may significantly extend a diver's safe exposure time in a high-temperature environment. Divers' subjective reports regarding the degree of hyperthermia they experience are very unreliable; we found that a diver can easily extend his stay in a high-temperature environment to the point of heat stroke. A positive-pressure, water-filled suit was developed, which was capable of maintaining a sufficient positive pressure within the suit to prevent leaks in case of puncture and to maintain appropriate body temperature.

Procedure

During the early phases of testing of polluted water diving apparatus by the NOAA Diving Program (NDP), hyperthermia became an obvious limiting factor to workload and dive duration at moderate temperatures (28°C). Diving suits normally limit or eliminate evaporative cooling of the skin. Breathing gas temperatures are approximately equal to water temperatures when using either scuba or surface-supplied diving equipment. Decontamination procedures following dives in contaminated water may require the diver to remain suited for up to 20 min, often in the hot sun.

A series of diving apparatus (USN MK 12 Helmet) integrity tests involved a 20-min period of light exercise, which stressed the potential failure points of the apparatus, and the more recently developed positive pressure water "suit under suit" (SUS).

Helmet, skin, and rectal temperatures of the divers were monitored continuously. Electrocardiographic (ECG) tracings were obtained during brief rest periods. Signals from the sensors were sent to the surface monitoring equipment via a shielded cable exiting through the diver's helmet. Initial tests were conducted in conjunction with integrity/decontamination studies at the NOAA Experimental Diving Unit. These tests involved diving in heated ammonia/fluorescent dye solutions in a 12-ft deep tank. The temperatures were increased on a daily basis until upper thermal limits were achieved.

Diver discomfort in air-filled suits became significant at temperatures in excess of 38°C. Tests were terminated after 20 min in 42°C water due to rapidly rising rectal temperatures and high heart rates. Rectal temperatures continued to rise after the removal of the diver from the water and during decontamination and undressing, suggesting an "afterise phenomenon."

During the same series, divers were able to complete three consecutive exercise periods in 44°C water using the SUS apparatus perfused with 22°C water. Hyperthermia was not limiting in these tests, and higher temperatures could not be obtained due to system heating limitations.

During the next series of tests, skin temperatures were measured on the chest, back, and crotch, as well as helmet and rectal temperatures and periodic ECG tracings. A heat exchanger was installed to provide temperature control of the cooling water.

In-line flow meters were used to determine water and air flow to the diver. A standard air flow rate of 6 ACFM was used throughout the test. Water flow was set at 7.5 liters per minute of 22°C water. The same 20-min light exercise series used in earlier tests were used in this new test series.

Initial tests in the 46°C range were uneventful. At 49°C, an unanticipated incident resulted in a significant overheating of one of the divers. The cornstarch used on the outside of the inner suit to allow easy entry into the outer suit became wet from contact with water between the two suits and turned into "gravy." The "gravy" plugged the screens on the ankle exhaust valves and prevented adequate perfusion of the lower portion of the suit, causing a rapid heating of the diver. Although he was uncomfortable, the diver was unaware of the extent of hyperthermia that he was experiencing. The diver was removed from the water and cooled with cold water towels around the head, chest, and shoulders. The following day the same diver successfully completed a 1.5 h dive in the 50.5°C water with a stable rectal temperature and elevated heart rate. On a subsequent dive an older diver (44 yr) exhibited a progressive increase in premature ventricular contractions (PVCs) as the dive progressed. The dive was terminated after 20 min.

Breathing a hot gas and having the head exposed to hot air while the body surface is maintained cool is an unusual situation for humans. The PVCs experienced by one diver and tachycardia by another caused sufficient concern to terminate the test series and to consider 50°C the maximum acceptable temperature for divers breathing ambient temperature gas with body cooling. Cooling of helmet gas may allow a significant increase in the maximum operational temperatures of diving.

We also noticed an increase in body temperature of surface tenders as they suited up in protective suits, breathing apparatus, and cooling vests. The increase in body temperature of surface personnel was relatively slow, up to a certain point, after which it would start to rise rapidly. This phenomenon was occurring among both the surface personnel and the divers. We identified this as the critical point, because once the individual's temperature started to rise rapidly, it took a significant time to cool him or to get him out of the adverse environment.

At least when working under the hot August sun, we could not maintain a stable temperature in the actively cooled tenders unless they stopped working and sat in the shade. We also observed that after they had unsuited and sat in the shade, where they obviously underwent a net body cooling, they still registered an increasing rectal temperature.

Conclusions and Recommendations

1. The highest recommended temperature for diving operations using helmets and air-filled suits is 40°C.

2. The highest recommended temperature for diving operations using helmets with ambient temperature breathing gases and body cooling is 50°C.

3. Divers' subjective feelings regarding their degree of hyperthermia are very unreliable, and a diver can easily extend his stay in hot water to the point of heat stroke. Reliable diver monitoring, to determine the degree of hyperthermia, is a recommended safety consideration when diving in heated water.

Physiology of Cold-Water Diving as Exemplified by Korean Women Divers

Y. S. PARK
Departments of Physiology
Kosin Medical College, Pusan, Korea,

SUK KI HONG
State University of New York at Buffalo
Buffalo, New York

Introduction

The specific heat of water is 1000 times and thermoconductivity 25 times greater than those of air. Consequently, the human body cools considerably faster in water than in air of the same temperature. This direct loss of body heat to the water is one of the most dominant problems of the diver even in tropical water, the temperature of which is still considerably lower than the thermoneutral level (34-35°C). Since 1960, we have been studying thermal physiology of Korean and Japanese professional breath-hold divers (both sexes) who engage in stressful diving work in cold water (22-25°C in summer and 10°C in mid-winter). Originally, we conducted extensive studies dealing with the assessment of the degree of cold stress and the pattern of cold acclimatization in traditional Korean women divers. More recently, these divers began to wear wetsuits (since 1977 in Korean divers), which gave us an opportunity to study the time course of deacclimatization to cold through a series of longitudinal studies. In addition, the insulative value of wetsuits while divers are engaged in underwater exercise has been critically reevaluated and some new important observations have been made.

Cotton-suit Divers

Figure 1A depicts the average time course of changes in the rectal temperature of 4 cotton-suit divers during a diving work shift in summer (23°C water) and winter (10°C water) (1). There were considerable individual variations, but in general the rectal temperature remained unchanged during the initial 5-10 min period, and then declined steadily to approximately 35°C after 30 min in winter and after 60 min in summer, at which time the divers voluntarily terminated diving work. The mean skin temperature dropped quickly from about 35°C to the level of water temperature; the reduction in mean body temperature was much greater in winter (8.4°C) than in summer (6°C). The calculated body heat debt was 28 and 119 kcal in summer and winter, respectively. These results indicate that the most important factor determining the working time is deep body cooling rather than the absolute amount of body heat loss.

Figure 1B illustrates the cumulative extra heat loss during a diving work shift (1). The extra heat loss is the sum of the extra heat production over the resting value and the change in body heat storage; thus it represents thermal cost of diving work. In both summer and winter the thermal cost increased rapidly during the initial period, followed by a steady, slow increase. However, the rate of the steady heat loss (slope of the steady portion of the curve) was much greater in winter (420 kcal/h) than in summer (175 kcal/h), due to more intensive shivering in the winter. The total diving heat loss during a work shift appeared to be 320 kcal in summer (60 min shift) and 480 kcal in winter (30 min shift). Since they took three shifts a day in summer and one or

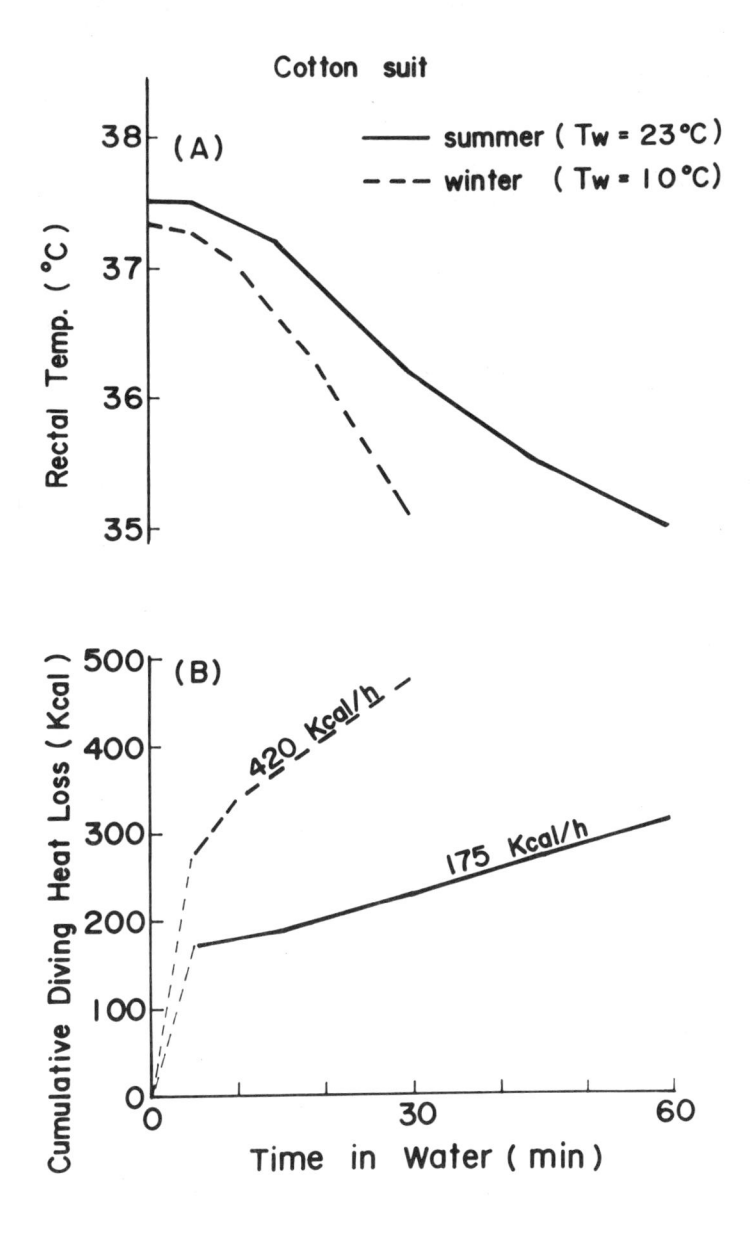

Fig. 1. Rectal temperature (Panel A) and the cumulative heat loss (Panel B) during diving work shifts in summer and winter. Redrawn from (1).

two in winter, the average daily diving heat loss was estimated to be of the order of 1000 kcal (2,3). A dietary survey indicated that the diver consumed approximately 3000 kcal/day in all seasons, which was 1000 kcal greater than that of the average Korean woman (2,3). To our knowledge, this magnitude of daily voluntary heat loss has never been observed in any other group of human subjects.

Since these cotton-suit divers subjected themselves to such a great cold water stress throughout the year, they appear to have acquired a unique type of cold acclimatization. Figure 2A shows that the basal metabolic rate of the diver increased markedly during the cold season (4,5). In nondiving Korean women of similar age and socioeconomic status there was no such seasonal variation in the basal metabolic rate. Since the protein intake, erythrocyte counts, and hemoglobin concentration were not different between the two groups, it was concluded that the increase in basal metabolic rate in divers in the cold season is a manifestation of a metabolic acclimatization to cold stress.

The shivering threshold of divers was much higher than that of nondivers (2,6,7). Figure 2B depicts the critical water temperatures of divers and nondivers as a function of subcutaneous fat thickness (6). The critical water temperature was defined as the lowest water temperature one can tolerate for 3 h without shivering; thus, a lower critical water temperature is synonymous with an elevated shivering threshold. The critical water temperature decreased with subcutaneous fat thickness, as depicted for subjects from a nondiving population. It is seen that divers who were very lean tolerated without shivering much cooler water than nondivers of comparable fat thickness. Such an elevation of shivering threshold in the diver was observed in both summer and winter.

As a consequence of the elevated shivering threshold, these traditional Korean women divers could maintain a higher maximal body insulation than nondivers (2,6,7). The maximal body insulation is linearly correlated with the subcutaneous fat thickness. Figure 2C compares the regression lines obtained from 30 divers and 89 nondivers (2). It is clear that, at a given subcutaneous fat thickness, the maximal body insulation is much greater in divers than in nondivers. The slope of the regression line is not different between the two groups, whereas the intercept with the y-axis is significantly higher in divers compared with nondivers, suggesting a higher nonfatty tissue insulation. These findings may indicate that divers are able to induce a more extensive vasoconstriction of the limb muscle or that divers possess a more effective countercurrent heat exchange mechanism in their limbs. Regardless of the mechanism, the basic pattern of this insulative acclimatization is similar to that observed in Australian aborigines (8).

Another feature of cold acclimatization observed in divers was a strong vasoconstriction in the most distal part of extremities during cold water immersion. Figure 2D illustrates the finger blood flow response to hand immersion in $6°C$ water (9). During 1 h immersion, divers maintained the finger blood flow at a significantly lower level than in nondivers, suggesting that the degree of vasoconstriction is significantly greater. Since the finger has a high surface area: mass ratio, such a reduction of blood flow would effectively reduce the heat loss during cold-water immersion.

Wet-suit Divers

To avoid severe cold water stress during diving work, Korean women divers began to wear wet suits in the 1970s. We therefore conducted a series of experiments to evaluate the effect of wearing wet suits on the thermoregulatory functions of these divers.

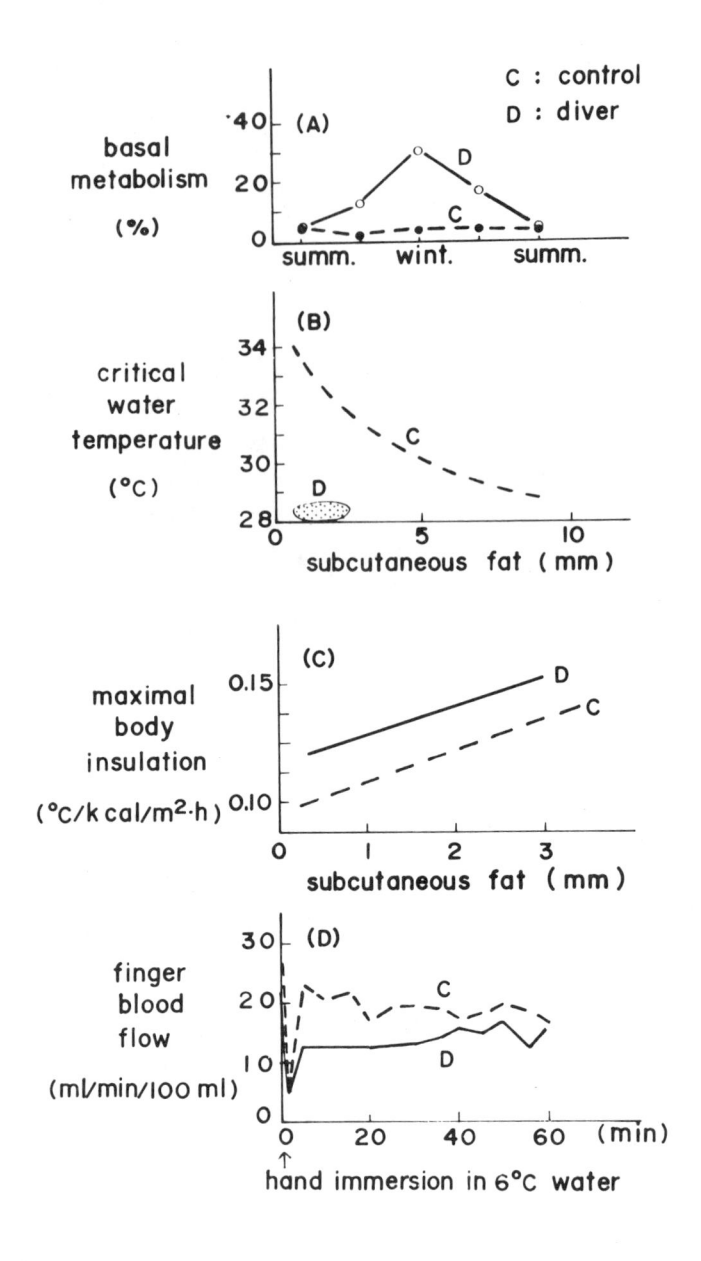

Fig. 2. Traditional cotton-suit divers compared with non-diving control subjects. Seasonal variations of basal metabolic rate (Panel A), critical water temperature (Panel B), and maximal body insulation (Panel C) as functions of subcutaneous fat thickness, as well as finger blood flow while the hand was immersed in 6°C water (Panel D). Data for panel A redrawn from (5), for panel B redrawn from (6), for panel C redrawn from (2), and for panel D redrawn from (9).

Thermal Exchanges During Diving Work

Figure 3 summarizes changes in the rectal temperature and the cumulative diving heat loss observed in 4 contemporary wet-suit divers (1). Unlike cotton-suit divers, the rectal temperature in wet-suit divers did not change appreciably over the 2-h work period. The total reduction of rectal temperature in 2 h was only 0.4°C in summer (23°C water) and 0.6°C in winter (10°C water); thus, the rectal temperature was of no major importance in the determination of work period. The mean skin temperature (31°C in summer; 28°C in winter) was also maintained at a level significantly higher than that in cotton-suit divers.

In both summer and winter, the cumulative diving heat loss increased steadily after the first 30 min in water. Thus, by extrapolating the steady portion of the curve we estimated that the modern protected divers, who usually engage in diving work for 3 h in a summer day and 2 h in a winter day, lose about 260 kcal in summer and 370 kcal in winter during a day. The rate of net heat loss, estimated from the slope of the steady portion of the curve, was approximately 80 kcal·h^{-1} in winter. These values are less than 35% of those observed in cotton-suit divers, indicating that the cold stress during diving work markedly decreased after divers adopted wet suits.

Deacclimitization to Cold Water Immersion

If the cold acclimatization observed in Korean women divers during cotton-suit era (Figure 2) was indeed developed through repeated exposures to severe cold water stress, it should disappear when the cold stress is removed by wearing wet suits.

Figure 4 summarizes the time course of deacclimatization of various thermoregulatory functions after wet suits were adopted in 1977 (10). The reversible increase in basal metabolic rate observed during cold seasons and the ability to maintain a high body insulation in cold water observed in both summer and winter had disappeared by 1980, i.e., within 3 years of wet suit diving. The mechanism of shivering suppression (i.e., a low critical water temperature) and the greater vasoconstriction of finger blood vessels during cold water immersion of a hand were sustained until year 3 of wet suit diving, but disappeared during the subsequent 3 years.

A winter-high and summer-low type of seasonal variation of basal metabolic rate has also been documented among Japanese divers (11), though the magnitude was much smaller than that observed in Korean divers. Interestingly, this variation of Japanese basal metabolic rate has been gradually diminished as the ratio of fat to carbohydrate (F:C ratio) in their diet has increased (12). Comparison of food surveys of contemporary divers (1) and traditional divers (3) indicated that the F:C ratio of Korean divers changed little over the last 20 years (0.097 in 1962 vs. 0.104 in 1982) so the lack of seasonal variations in the basal metabolic rate among modern divers cannot be attributed to dietary changes. The elevated basal metabolism observed in previous divers during cold seasons appears to be a manifestation of a metabolic acclimatization to cold.

The attenuation of shivering mechanisms observed in the cotton-suit divers has been interpreted as an acclimatization process (2) and the mechanism underlying the shivering suppression is not clearly understood. However, our studies showed that, when the shivering threshold of divers was higher than that of nondivers (such as in 1980), divers began to shiver at a lower core temperature than nondivers at the same cold water stress to the skin. During cold water immersion divers also usually complained of internal chilling, not external chilling, whereas nondivers complained of external chilling (10). These facts may imply that the sensitivity of cutaneous cold receptors may be suppressed in divers, as has been observed in animals which are acclimatized to long-term cold (13).

A relatively high maximal body insulation in traditional cotton-suit divers was due to a higher insulation of the nonfatty shell than in nondivers (Fig. 2C). Since the critical water temperature of divers was lower than that of nondivers (Fig. 2B), passive cooling of the peripheral tissues (including muscles) might be

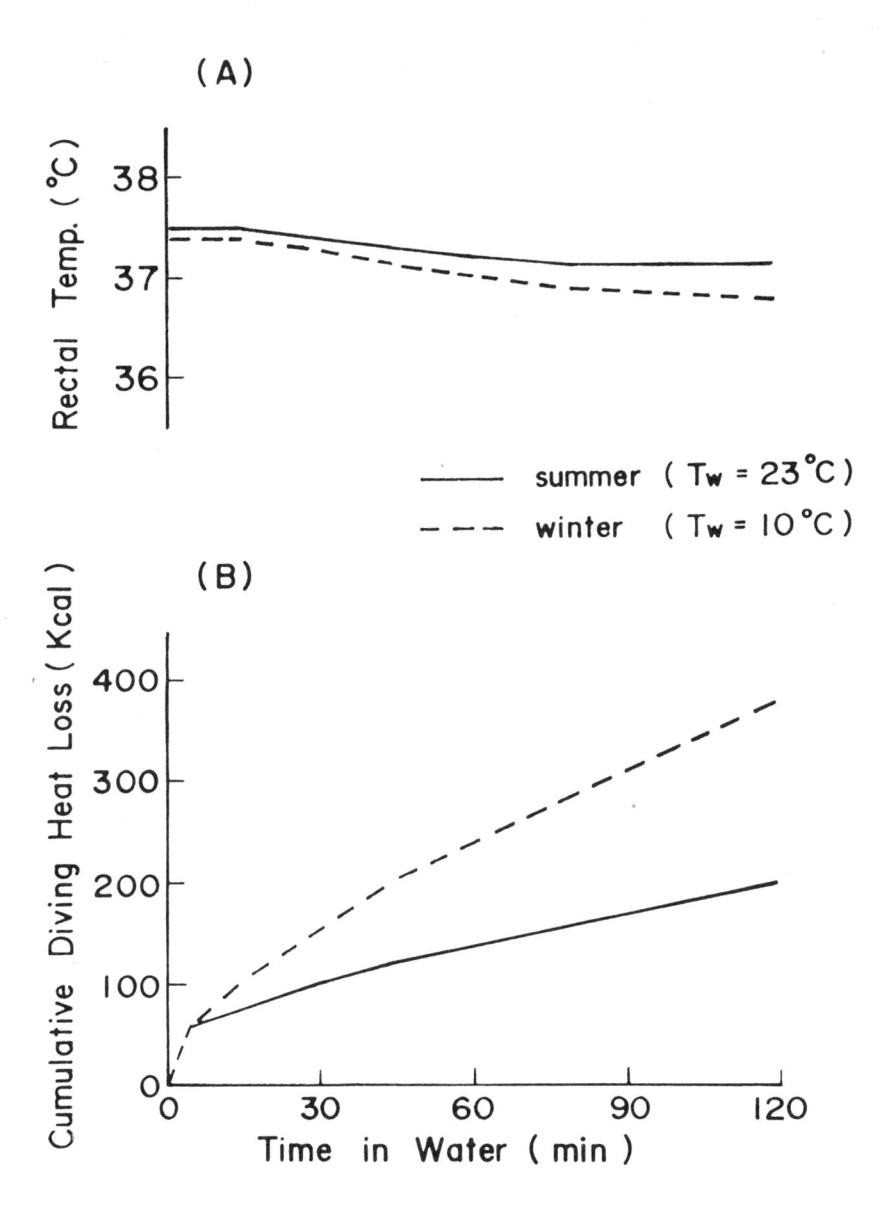

Fig. 3. Rectal temperature (Panel A) and cumulative heat loss during work in contemporary wet-suit divers ; redrawn from (1).

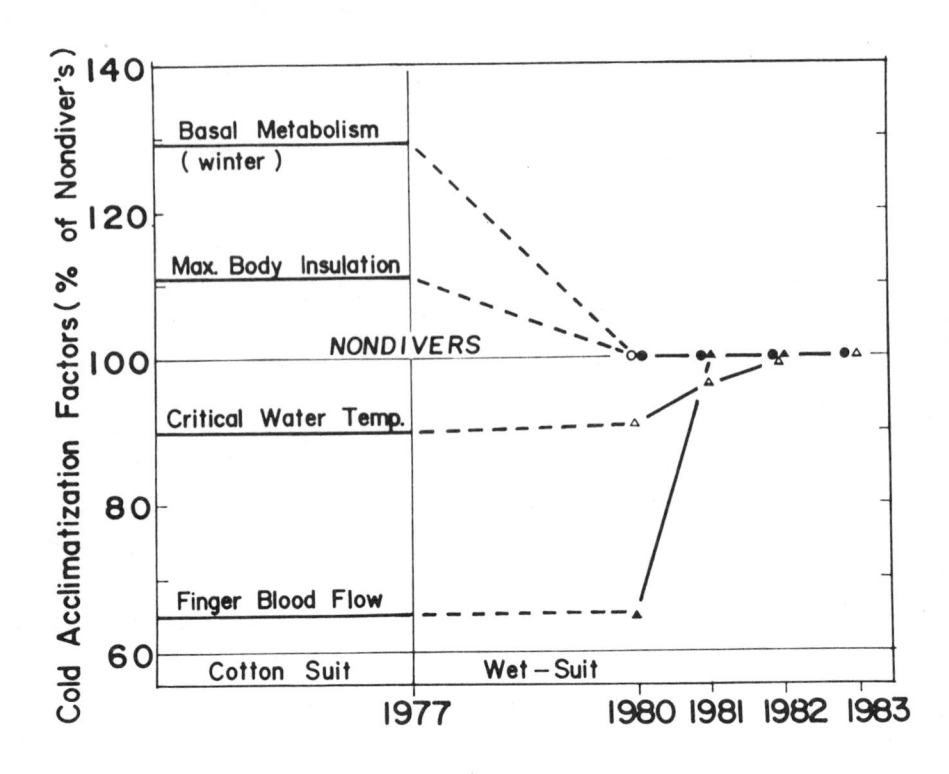

Fig 4. *Time course of disappearance of various acclimatization parameters since the adoption of wet suits in 1977; redrawn from (10).*

greater, providing a thicker layer of nonfatty shell. In other words, the greater maximal body insulation observed in divers could be secondary to the elevation of their shivering threshold. However, the fact that the insulative acclimatization of the peripheral tissue disappeared in contemporary divers even faster than the mechanism of shivering attenuation suggests strongly that the elevation of the maximal body insulation is distinct from the elevation of shivering threshold. We therefore concluded that the insulative acclimatization of the peripheral tissue reflects a vascular acclimatization in which the diver could induce either a more extensive vasoconstriction of the limb musculature or a more effective countercurrent heat exchange in the limb (10).

Finally, reversal of the more intense vasoconstriction of the finger vessels during hand immersion in cold water took 4 years. Perhaps this delayed return of the finger vascular response to normal was because the cold water stress to most parts of the body was immediately reduced by wearing wetsuits but the cold stress to hands was not. Contemporary divers do not wear protective gloves even in winter so their hands are still subjected to cold stress. This in turn suggests that the local vascular response observed in traditional divers had not been developed through the cold water stress to hands, but through the stress to the whole body. In this connection, it is important to point out that Gaspe fisherman (14) and British fish-filleters (15), who routinely expose their bare hands to cold water, have a considerably weaker finger vascular constriction during hand immersion than control subjects. Evidently, this type of acclimatization to local cold exposure is basically different from that of whole-body cold exposure.

Thermal Insulation

The reduction of diving heat loss in Korean women divers after adoption of wet suits was apparently due to additional insulation provided by the suit. The overall thermal insulation (I_{total}) estimated in four divers using values of rectal-to-water temperature difference ($T_{re}-T_w$) and skin heat flux (H_{sk}) during diving work ($I_{total} = [T_{re}-T_w]/H_{sk}$) was approximately 2.5 times greater in wet suit divers (0.170-0.193°C· kcal^{-1} m^{-2}·h^{-1}) than in cotton-suit divers (0.068-0.081) (1). Since, however, the physical insulation of the wet suit may vary with the depth of diving due to compression of the trapped air in the suit (16), and since the physiologic insulation of the body may also decrease with exercise due to an increase in blood circulation to the working muscle (17), the overall insulation of the wet suit diver will change with depth and workload.

Figure 5 depicts the effect of exercise on the steady-state insulation in nude (cotton-suited) (A) and wet-suited (B) subjects immersed in water of critical temperature (17). The subjects were immersed up to the neck in a circulating water bath and either rested for 3 h or exercised at a constant intensity using a submerged bicycle ergometer. As shown in Figure 5A, the body insulation declined progressively as the exercise intensity (expressed as a metabolic rate above resting value) increased. On the average, the insulation decreased to approximately 50% of the resting value with exercise at 2 Met (1 Met = 50 kcal · m^{-2} h^{-1}). Thus, the skeletal muscle appears to provide as much as 80% of the total body insulation, with the subcutaneous fat and skin accounting for the remainder. This suggests that, for subjects immersed in cold water, the heat loss is controlled largely by blood flow to the skeletal muscle, with the subcutaneous fat and skin playing a less important role than commonly supposed

The reason the skeletal muscle plays such an important role in providing insulation is that thermal control of blood flow, and hence insulation, is mostly accomplished in the extremities and not in the trunk (18). Cannon and Keatinge (19) have observed that, in the subject immersed in water, the thermal conductance of the hands and feet decreased drastically from 0.048 cal·cm^{-2}·min^{-1}·C^{-1} °C at 35°C to 0.001 at 22°C. Over the same temperature range, the conductance of the chest decreased by only 36% from 0.028 to 0.018 cal·cm^{-2}·min^{-1}·°C^{-1} Barcroft and Edholm (20) observed that, when an arm is immersed in 13°C water, the forearm blood flow falls to only 0.5 ml ·100 ml^{-1}·min^{-1} compared with 17.6 ml·100 ml^{-1}

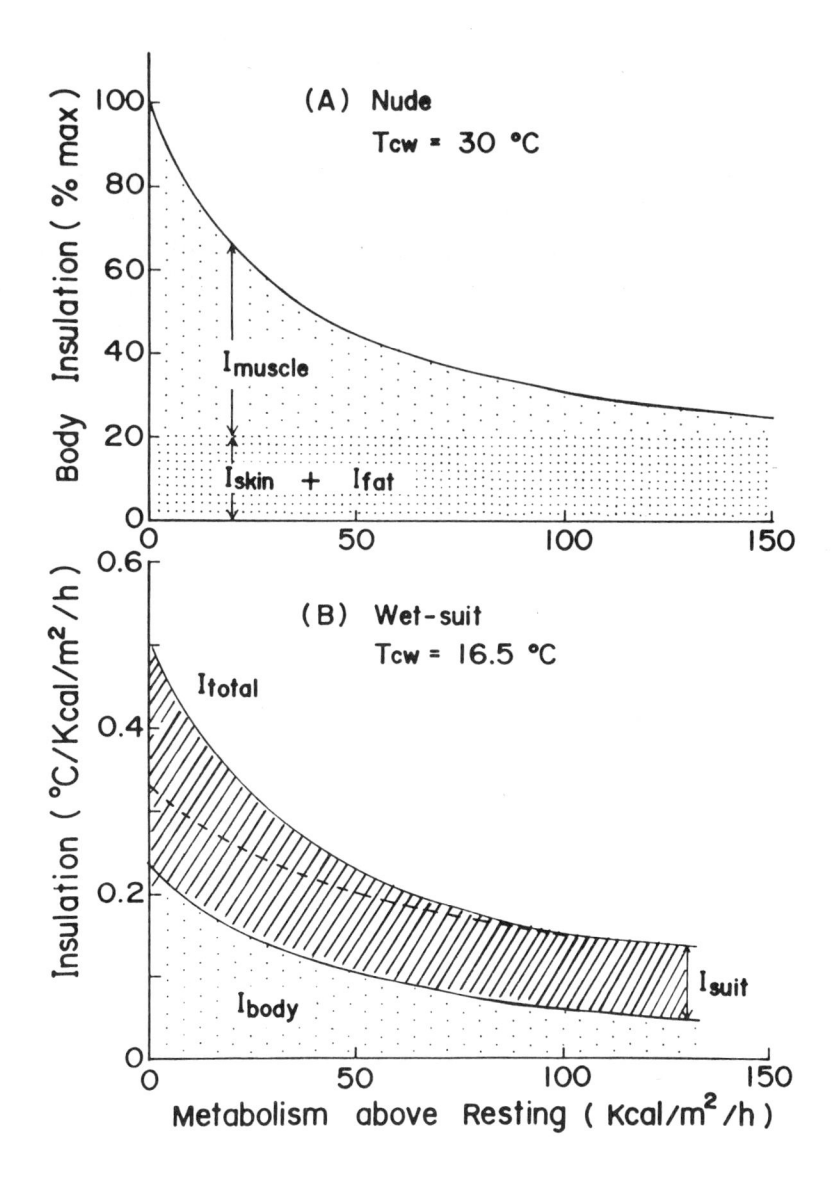

Fig. 5. Thermal insulations as function of metabolism above the resting level in water of critical temperature; cotton-suited (nude) in Panel A, redrawn from (17) and wet-suited in Panel B, redrawn from (23).

min^{-1} observed during immersion in 43°C water. Such a reduction in limb blood flow will not only retard heat transfer from the body core to the extremities but will also increase the efficiency of countercurrent heat conservation mechanisms, which are effective only when blood flow is sufficiently small (21). Therefore, most of the heat generated in the core of the subject while resting in cold water is lost through the trunk surface rather than through the limbs.

The counter-current mechanism may be even more important in wet-suited subjects. Since the physical insulation of foamed neoprene will decrease as the curvature of surface increases (22), the insulative value of wet suits will be much smaller in the limb than in the trunk. Furthermore, the design of diver wet suits is such that most of the trunk surface is covered by double sheets (pants and jacket) and the limbs by a single sheet. Thus, wet suits provide a good insulation to the trunk but a poor insulation to the limbs. Consequently, a restriction of blood flow to the limbs should greatly increase the efficiency of thermoregulation in the wet suited diver. We have tested this hypothesis in a study on Korean women wet-suited divers (23).

As shown in Fig. 5B, the overall (total) insulation decreased from about 0.5°C·kcal^{-1}·m^{-2}·h^{-1} at rest to approximately one half at 2 Met and to one third of the resting value at 3 Met exercise. This decrease in total insulation (I_{total}) appeared to be due in part to the reduction in body insulation (I_{body}) and in part to the decrease in insulation afforded by wet suits (I_{suit}). The apparent I_{suit} estimated from the difference between the I_{total} and I_{body} (striped area in Fig. 5B) was, on the average, 0.27°C·kcal^{-1}·m^{-2}·h^{-1} at rest, but it decreased gradually with exercise intensity until it reached approximately 0·12°C·kcal·m^{-1}·h at 3 Met or above, similar to the physical insulation of 5-mm neoprene wet suit obtained using a copper manikin (24). Physical insulation of the wet suit should not change unless its thickness is changed. Since there was no apparent reason to expect a difference in the suit thickness between rest and exercise, we speculated that the unexpectedly high functional insulation of the wet suits in resting subjects is a consequence of physiologic regulation in cold water.

The skin temperature underneath the wet suit will become much lower in the extremities than in the trunk during immersion in cold water, so immersion with wet suits is analogous to exposing the limb to water colder than that exposing the trunk. This will lead to a strong vasoconstriction in the extremities. Restriction of limb blood flow will greatly reduce the surface area for heat exchange, and most of the heat exchange between the body core and water will take place at the trunk surface where the suit insulation is relatively high (double sheets and low curvature). For these reasons, wet suits provide far greater physiologic insulation at rest than during exercise. Exercise hyperemia reduces not only the thermal insulation from the deep tissue to the skin but also the thermal insulation down the length of the limb, so much of the heat produced in the skeletal muscle is dissipated through the large surface area of limbs rather than returning to the body core. Figure 6 shows the effective heat exchange area A estimated in four Korean women wet-suit divers using the following formula (23): $A = I_{suit} \times 0.92 M/(T_{sk} - T_w)$, where M is the steady-state metabolic rate in $kcal \cdot h^{-1}$, T_{sk} is the steady-state mean skin temperature, and I_{suit} is the physical insulation of wet suits, which was assumed to be constant at $0.12°C \cdot kcal^{-1} \cdot m^{-2} \cdot h^{-1}$ in all conditions. The area at rest was only 0.55 m^2, which was equivalent to 40% of the total suit surface area, but it increased with exercise and became identical to the actual suit surface area at a metabolic rate of approximately 200 $kcal \cdot h^{-1}$. This analysis strongly supports a notion that the relatively high apparent suit insulation in the resting subject may be due to a reduced surface area for heat exchange. One obvious practical implication of these findings is that if a wet-suited diver is in a situation where escape from the cold water is not possible, it is better to hold still than to swim if wasting of energy is to be prevented.

Figure 7 illustrates the effect of hydrostatic pressure on the thermal insulation in wet-suit divers (25). Subjects were im-

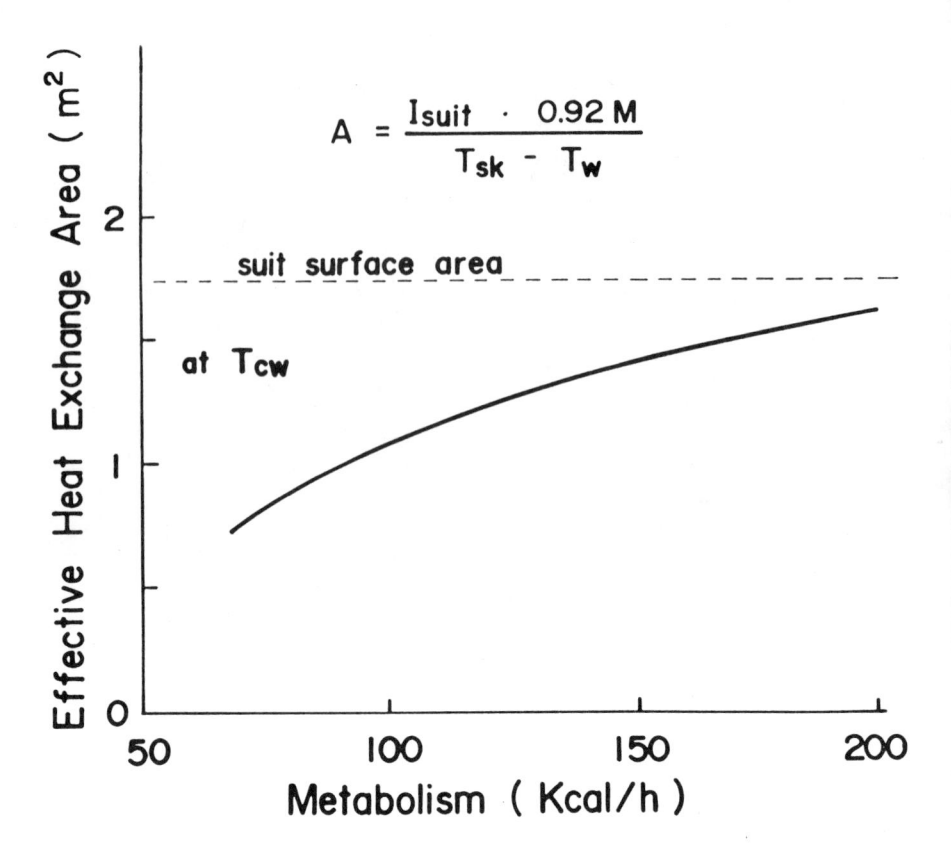

Fig. 6. Predicted changes in effective heat exchange surface area as function of metabolic rate in wet-suited divers. Metabolic rate was altered by adjusting the level of exercise in water of critical temperature; redrawn from (23).

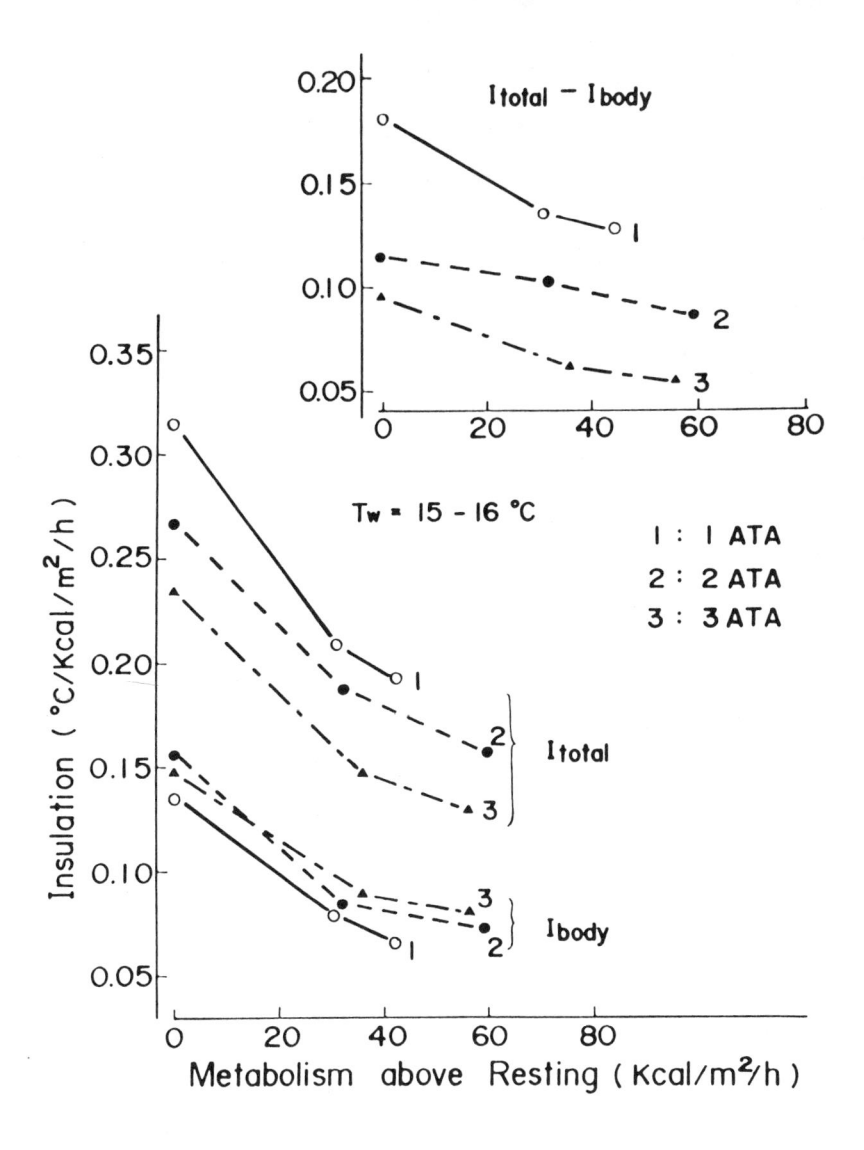

Fig. 7. Total, body and wet-suit insulation as function of the metabolism at 1, 2, and 3 ATA of air pressure in wet-suit divers; redrawn from (25).

mersed up to the neck for 2-3 h in 15-16°C water in the wet-pot of a hyperbaric chamber. The chamber pressure was maintained at 1, 2, or 3 atm abs air. At all pressures, insulations declined inversely with exercise intensity. The I_{total}, either at rest or during exercise, decreased as the pressure increased. On the other hand, the I_{body} increased slightly at pressure than at the surface (1 atm abs). Consequently, the apparent I_{suit} (which is the difference between I_{total} and I_{body}) decreased as the pressure increased (inset). The actual suit thickness decreased from 5-mm at 1 atm abs to 3.5 at 2 and 2.6 at 3 atm abs, so the physical insulation of the suit was decreased with pressure, and consequently, skin surface under the suit was cooled more as the pressure increased. The mean skin temperature at rest was 27°C at 1 atm abs, < 24.5°C at 2 atm abs, and 24.2°C at 3 atm abs. The relatively high I_{body} at 2 and 3 atm abs, as compared with 1 may be attributed to more intensive peripheral vasoconstriction induced by the lower skin temperature at pressure.

In breath-hold diving, divers repeat the cycle of a dive and surface recovery, each lasting for 30-40 sec in the case of Korean women divers. During surface recovery, divers are resting in a state of head-out water immersion, whereas during diving they exercise and are exposed to various degrees of hydrostatic pressure. Since the body insulation changes with exercise, and the suit insulation decreases with pressure, as described above, the overall insulation of a breath-hold wet-suit diver will increase as the ratio of surface to dive time increases. Figure 8 summarizes the effect of voluntarily changing the surface to dive time ratio from 1 to 2 on the thermal insulation of two Korean women wet-suit divers. The average I_{body} increased from 0.062 to 0.075°C·kcal·m^2·h and the I_{suit} from 0.126 to 0.153°C·kcal^{-1}·m^2·h^{-1}; hence, the overall insulation increased by about 23%. Thus, by adjusting the surface to dive time ratio one can prolong the working time without increasing heat loss during breath-hold diving. This wisdom of behavioral adjustment of diving pattern is actually observed in

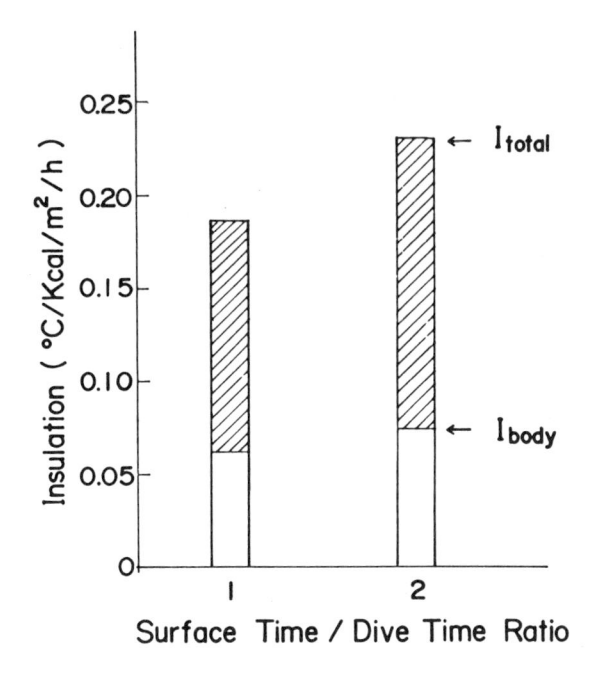

Fig. 8. Total and body insulation of wet-suit, breath-hold divers as function of the ratio of surface time to dive time from two women who dove to 4 or 5 meters in summer; redrawn from (26).

male wet-suit divers of Tsushima Island of Japan whose surface to dive time ratio changes from 1.30 in summer to 1.97 in winter (26).

Effect of Wearing Gloves

Since contemporary Korean women divers do not wear protective gloves even during winter, we have evaluated the effect of wearing gloves on their thermal balance during immersion in cold water (27). Subjects, clad with 5-6-mm-thick wet suits (jacket, pants, and boots) with or without wearing 3-mm-thick neoprene gloves, were immersed in water of critical temperature (17.3°C) and remained still for 3 h. Table 1 shows that the rectal temperature was slightly but significantly lower with gloves, and the skin heat flux was significantly higher. In both hands and forearms, the regional heat flux determined directly with a heat flux transducer was higher and the thermal insulation index, calculated by dividing the rectal-to-local skin temperature difference with the local skin heat flux, was lower with gloves than without gloves (data not shown). Although gloves provided some additional insulation, their value was relatively so small that the total local insulation was still significantly lower than the tissue insulation alone without gloves. These results indicate that in wet-suited subjects resting in cold water, gloves do not provide additional protection against heat loss, but rather decrease the efficiency of thermoregulatory mechanisms. The mechanism for this phenomenon is not entirely clear, but because overall insulation of the body is mostly determined by the longitudinal regulation down the length of the limbs rather than the local insulation, we speculate that blood flow to the extremities was more effectively reduced by direct exposure of hands to cold water. Regardless of the mechanism, these findings suggest strongly that sensory inputs from the cold receptors in the distal extremities is particularly important in thermoregulation during immersion in cold water.

Table 1. Effects of gloves during immersion at critical water temperature[a].

	No gloves	Gloves
Rectal temperature (°C)	37.1	36.8[b]
Skin temperature (°C)	28.5	27.9
Metabolic rate (kcal·m^2·h^{-1})	44.8	49.0
Skin heat flux (kcal·m^2 h^{-1})	46.7	54.7[b]
Total insulation (°C·kcal·m^2·h^{-1})	0.43	0.36

[a]From Choi et al. (27).
[b]Significantly different from the corresponding value without gloves.

References

1. Kang, D.H., Y.S. Park, Y.D. Park, et al. 1983. Energetics of wet-suit diving in Korean women breath-hold divers. J. Appl. Physiol. 54:1702-1707.

2. Hong, S.K. 1973. Pattern of cold adaptation in women divers of Korea (ama). Fed. Proc. 32:1614-1622.

3. Kang, D.H., P.K. Kim, B.S. Kang, S.H. Song, S.K. Hong. 1965. Energy metabolism and body temperature of ama. J. Appl. Physiol. 20:46-50.

4. Hong, S.K. 1963. Comparison of diving and nondiving women in Korea. Fed Proc 22:831-833.

5. Kang, B.S., S.H. Song, C.S. Suh, S.K. Hong. 1963. Changes in body temperature and basal metabolic rate of the ama. J. Appl. Physiol. 18:483-488.

6. Rennie, D.W. 1963. Thermal insulation of Korean diving women and nondivers in water. pp. 315-324. In: H. Rahn (ed.) Physiology of Breath-hold Diving and the Ama of Japan. National Academy of Sciences, Washington, DC.

7. Rennie, D.W., B.G. Covino, B.J. Howell, S.H. Song, B.S. Kang, S.K. Hong. 1962. Physical insulation of Korean diving women. J. Appl. Physiol. 17:961-966.

8. Hammel, H.T., R.W. Elsner, D.H. Le Messurier, M.T. Anderson, F.A. Milan. 1959. Thermal and metabolic responses of the Australian aborigine to moderate cold in summer. J. Appl. Physiol. 14:605-615.

9. Paik, K.S., B.S. Kang, D.S. Han, D.W. Rennie, S.K. Hong. 1972. Vascular response of Korean ama to hand immersion in cold water. J. Appl. Physiol. 32:446-460.

10. Park, Y.S., D.W. Rennie, I.S. Lee, et al. 1983. Time course of deacclimatization to cold water immersion in Korean women divers. J. Appl. Physiol. 54:1708-1716.

11. Yurugi, R., T. Sasaki, H. Yoshimura. 1972. Seasonal variation of basal metabolism in Japanese. pp. 395-410. In: S. Itoh, K. Ogata, H. Yoshimura (eds.) Advances in Climatic Physiology. Igakushoin, Tokyo.

12. Yurugi, R., H. Yoshimura. 1975. Seasonal variation of basal metabolism in Japanese. pp. 45-49. In: H. Yoshimua, S. Kobayashi (eds.) Physiological Adaptation and Nutritional Status of the Japanese. University of Tokyo Press, Tokyo.

13. Hensel, H. 1981. Neural processes in long-term thermal adaptation. Fed. Proc. 40:2830-2834.

14. Le Blanc, J. 1962. Local adaptation to cold of Gaspe fisherman. J. Appl. Physiol. 17:950-952.

15. Nelms, J.D., D.J.G. Soper. 1962. Cold vasodilation and cold acclimatization in the hands of British fish-filleters. J. Appl. Physiol. 17:444-448.

16. Beckman, E.L. 1967. Thermal protective suits for underwater swimmers. Mil. Med. 132-195-209.

17. Park, Y.S., D.R. Pendergast, D.W. Rennie. 1984. Decrease in body insulation with exercise in cool water. Undersea Biomed Res. 11:159-168.

18. Burton, A.C., O.G. Edholm. 1969. Man in a Cold Environment. Hafner, New York.

19. Cannon, P., W.R. Keatinge. 1960. The metabolic rate and heat loss of fat and thin men in heat balance in cold and warm water. J. Physiol. (Lond) 154:329-344.

20. Barcroft, H., O.G. Edholm. 1943. The effect of temperature on blood flow and deep body temperature in the human forearm. J. Physiol. (Lond) 102:5-20.

21. Bazett, H.C., L. Love, M. Newton, L. Eisenberg, R. Day, R. Forster. 1948. Temperature changes in blood flowing in arteries and veins in men. J. Appl. Physiol. 1:3-19.

22. Van Dilla, M., R. Day, P.A. Siple. 1948. Special problem of hands. pp. 374-388. In: L.H. Newburgh (ed.) Physiology of Heat Regulation and the Clothing. Saunders, Philadelphia.

23. Yeon, D.S., Y.S. Park, J.K. Choi, et al. 1987. Changes in thermal insulation during underwater exercise in Korean female wetsuit divers. J. Appl. Physiol. 62:1014-1019.

24. Goldman, R.F., J.R. Breckenridge, E. Reeves, E.L. Beckman. 1966. "Wet" versus "dry" suit approaches to water immersion protective clothing. Aerosp. Med. 37:485-487.

25. Suh, D.J., D.S. Yeon, H.J. Kim, et al. 1987. Thermal balance of wetsuit divers during exercise in cold water at 1, 2, and 3 ATA. pp. 121-129. In: A.A. Bove, A.J. Bachrach, L.J. Greenbaum (eds.) Underwater Physiology IX. Proceedings of the Ninth Symposium on Underwater and Hyperbaric Physiology. Undersea and Hyperbaric Medical Society, Bethesda, MD.

26. Shiraki, K., N. Konda, S. Sagawa, Y.S. Park. 1987. Diving pattern and thermoregulatory responses of male and female wet-suit divers. pp. 124-133. In: C.E.G. Lundgren, M. Ferrigno (eds.) The Physiology of Breath-hold Diving. Undersea and Hyperbaric Medical Society, Bethesda, MD.

27. Choi, J.K., Y.S. Park, J.S. Kim, et al. 1988. Effect of wearing gloves on the thermal balance of Korean women wet-suit divers in cold water. Undersea Biomed. Res. 15:155-164.

The Liquid-filled Suit/Intersuit Concept: Passive Thermal Protection for Divers

M. L. NUCKOLS, M. W. LIPPITT
Naval Systems Engineering
US Naval Academy
Annapolis, Maryland

J. DUDINSKY
Naval Coastal Systems Center
Panama City, Florida

Introduction

The liquid-filled suit/intersuit concept is intended to provide passive thermal protection for cold underwater missions of long duration. Conventional passive approaches use micro-fibrous batts beneath lightweight drysuits. The insulation thickness to maintain thermal comfort during long missions at low metabolic levels would be excessively bulky and overly buoyant for subsequent swimming scenarios.

The liquid-filled suit/intersuit can provide the resting diver with a liquid layer having a density approximately that of water and low thermal conductivity. This results in added insulation without the additional buoyancy and bubble migration to shoulders and neck region that occurs when inflating drysuits with a gas. When the diver is required to swim from a free-flooding submersible, the liquid can be drained from the interlayer to reduce the insulation so that the diver is essentially swimming in a conventional drysuit with a Thinsulate undergarment (Thinsulate is a registered trade name used by 3M corporation for its micro-fibrous polypropylene material). This concept can also be beneficial during extended in-water decompressions for deep salvage missions; as the diver rests during a decompression stage, the suit/intersuit can be inflated with insulating fluid.

The primary advantage of the liquid-filled suit is that the diver can be protected from the cold without an active heating source. A beneficial side effect is that the insulating liquid is unaffected by suit squeeze, giving the feet and legs further protection. By selecting an insulating liquid which has a specific weight approximately that of water, minimal buoyancy variations will occur as the fluid level in the suit is varied. Table 1 lists some potential fluids. Toxicologic and suit-material compatibility concerns are very important, but are not addressed in this analysis.

Thermal Insulation Potentials

The liquid-filled suit/intersuit consists of a Passive Diver Thermal Protection System (PDTPS) developed by the US Navy, having a B400 Thinsulate undergarment and a tri-laminate drysuit, covered with an elastic outer drysuit. Previous testing (1) has characterized the insulation value of the PDTPS as 1.2 clo (0.95 BTU/ft$^2 \cdot$h$\cdot$°F). The outer drysuit is assumed to be a 0.030-inch-thick, reinforced elastomer (K = 0.12 BTU/ft$\cdot$h°F). The reciprocal of the total thermal conductance (H$_{TOTAL}$) of the liquid-filled suit/intersuit can be approximated as

$$1/H_{TOTAL} = \frac{1}{\dfrac{1}{H_{PDTPS}} + \dfrac{X_{OG}}{K_{OG}} + \dfrac{X_{LIQ}}{K_{LIQ}}}$$

Table 1. Liquids with low thermal conductivity.

Liquid	Thermal Conductivity, BTU/ft^{-1}·h^{-1}·°F^{-1}	Density lb/f^3
Fluorinert FC-77[a]	0.037	110.0
Light Oil	0.077	57.0
Aniline	0.100	63.8
n-Butyl Alcohol	0.089	50.6
Ethylene Glycol	0.100	69.2
Glycerine	0.133	78.7
Isobutyl Alcohol	0.082	50.0
Kerosene	0.086	51.2
Turpentine	0.073	53.9
Propylene Glycol	0.116	65.5
Ethyl Alcohol	0.097	49.2
Heptyl Alcohol	0.094	51.2
Pentane	0.078	39.1
Water[b]	0.348	62.3

[a]Fluorinert is an electronic liquid trade name.
[b]Water is shown for comparison.

where
>X is the layer thickness, ft
>K is the thermal conductivity, $BTU/ft^{-1} \cdot h^{-1} \cdot °F^{-1}$
>H is the thermal conductance, $BTU/ft^{-2} \cdot h^{-1} \cdot °F^{-1}$.

Subscripts are
>OG, outer garment drysuit
>LIQ, liquid layer
>PDTPS, passive diver thermal protection system.

$$1/H_{TOTAL} = \frac{1}{\frac{1}{0.95} + \frac{0.3/12}{0.12} + \frac{X_{LIQ}}{K_{LIQ}}} = \frac{1}{1.073 + \frac{X_{LIQ}}{K_{LIQ}}}$$

The effective suit insulation, expressed in units of clo can then be written

$$\text{Effective Suit clo} = \frac{1.136}{H_{TOTAL}} = 1.22 + 1.136 \frac{X_{LIQ}}{K_{LIQ}}$$

Estimates of the mission durations that would be permissible with the liquid-filled suit can be derived from an energy balance for a diver as follows:

$$\dot{S} = \dot{Q} + \dot{M} - (\dot{Q}_{RESP} + \dot{C}_S + \dot{R} + \dot{W}) \quad [1]$$

where

>$\dot{Q}$ = supplemental heating (active)
>$\dot{M}$ = metabolic heat production
>$\dot{Q}_{RESP}$ = respiratory heat loss
>$\dot{C}_S$ = suit heat loss
>$\dot{R}$ = radiation heat loss,
>$\dot{W}$ = diver work rate, and
>$\dot{S}$ = rate of energy storage/loss from diver.

It is assumed that

$\dot{Q} = 0$, since concept depends on passive insulation only;

$\dot{W} = 0$, since the diver is resting; and

$\dot{R} = 0$, since radiation heat loss is negligible under water.

Based on the Bureau of Medicine's (BUMED) thermal criteria for cold water diving (2), the dive should be terminated when $S/m = -5.4$ BTU·lb, where m is the mass of the diver's body and S is the lost body heat. Since $\dot{S} = S/t$, where t is the mission time, then $\dot{S} = -5.4\, m/t$. By substituting the above assumptions and equalities into Equation [1], we obtain

$$(\dot{M} - \dot{Q}_{RESP}) - \dot{C}_S = -5.4\, m/t$$

But

$$\dot{C}_S = H A (\overline{T}_{MS} - T_\infty) = (1.136/\text{clo}) A (\overline{T}_{MS} - T_\infty)$$

where H is the overall convective heat transfer coefficient between the diver and the water, A is the body surface area, $\overline{T}_{MS}$ is the mean skin temperature of the diver, and T_∞ is the surrounding water temperature.

If we assume the "average diver" is 178 lbs (3) and has a surface area of 19.4 ft^2, with a minimum mean skin temperature given by BUMED (2) as 77°F (this assumption will give conservatively high estimates of suit heat loss as the diver becomes chilled), then we can solve [Eq. 1] for the permissible mission duration as

$$t, \text{hrs} = -961/(\dot{M} - \dot{Q}_{RESP}) - 22.04\,(77 - T_\infty)/\text{clo}) \quad [2]$$

For shallow depths $\dot{Q}_{RESP}$ is small compared to $\dot{M}$ (approximately 5% at the surface). Therefore, for this analysis estimates for $\dot{M} - \dot{Q}_{RESP}$ will be assumed to be the metabolic level for a resting diver in cold water: 450 BTU·hr. Table 2 shows the estimated mission durations from [Eq. 2]. Acceptable durations in excess of 6 hours are indicated with only a 0.5-inch liquid layer thickness in 28°F water; this compares with approximately 2 hours when using the PDTPS. This improvement is further magnified in 35°F water; 17 hour durations are indicated with a 0.5-inch liquid layer, whereas, only 3 hours are allowable with the PDTPS.

It is possible that localized active heating would still be necessary when using liquid-filled suits in long, cold missions, but the power storage requirements would be only a fraction of that needed with current active heating of the whole diver.

References

1. Lippitt, M.W., M.L. Nuckols. 1982. The development of an improved suit system for cold water diving. NCSC Report TM-336-82, Naval Coastal Systems Center, Panama City, FL 32407.

2. Beatty, H.T., T.E. Berghage. Diver Anthropometrics, NEDU Report 10-72, Naval Experimental Diving Unit, Panama City, FL 32407.

3. Webb, P., E.L. Beckman, P.G. Sexton, W.S. Vaughan. 1976. Proposed thermal limits for divers: a guide for designers of thermally protective equipment. ONR Contract N00014-72-C-0057.

Acknowledgment

The authors express their sincere appreciation to Mr. Martin Harrell, Eastport International, Inc.

Table 2. Estimated mission durations.

Liquid thickness inches	clo	T, °F	Time, hrs $\dot{M} - \dot{Q}_{RESP} = 350$	$\dot{M} - \dot{Q}_{RESP} = 450$
0	1.22	28	1.8	2.2
		35	2.4	3.1
		45	4.2	7.5
		55	20.3	> 24
0.25	1.53	28	2.7	3.8
		35	3.8	6.2
		45	8.7	> 24
		55	> 24	> 24
0.5	1.83	28	4.0	6.9
		35	6.2	17.2
		45	> 24	> 24
		55	> 24	> 24
0.75	2.14	28	6.2	17.6
		35	11.6	> 24
		40	> 24	> 24
1.0	2.45	28	10.6	> 24
		35	> 24	> 24

> = more than

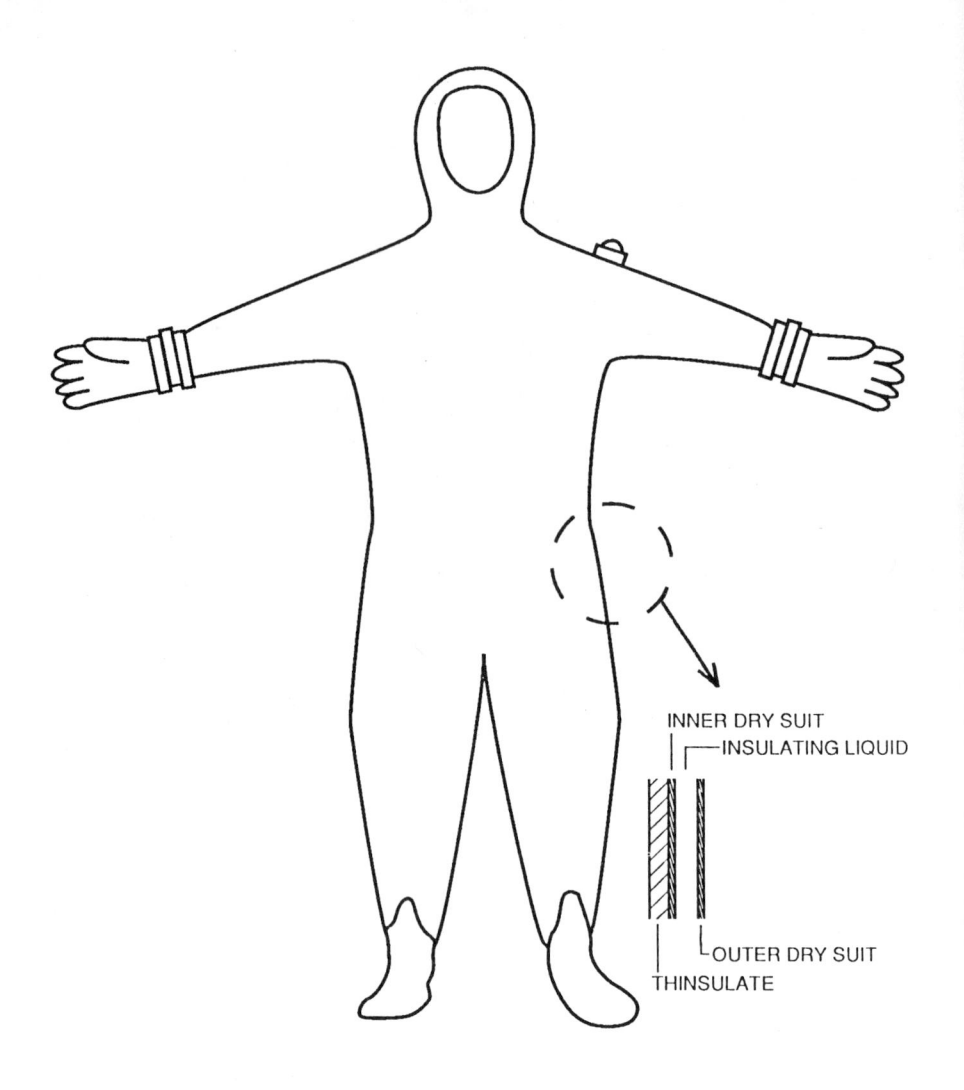

Figure 1. Liquid-filled suit/intersuit concept

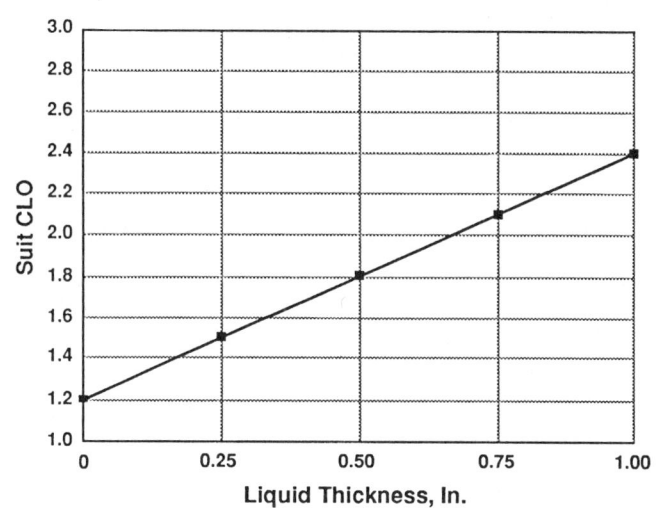

Figure 2. Liquid-filled suit/intersuit insulation potentials.

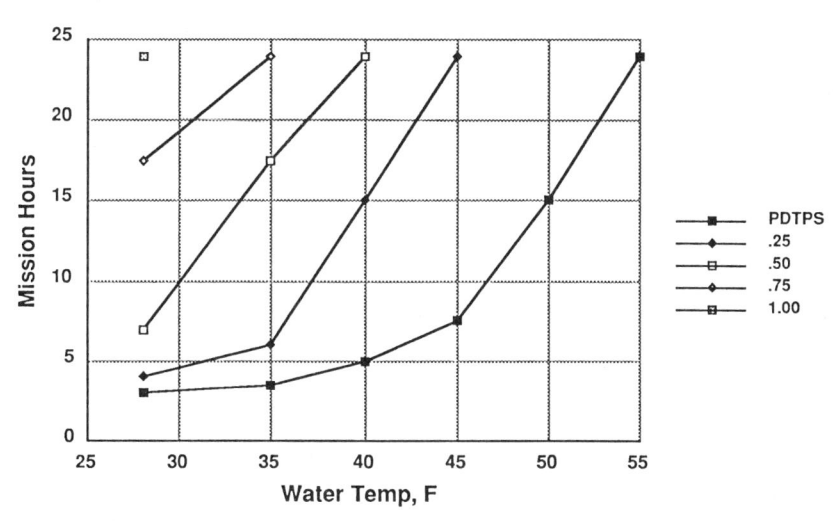

Figure 3. Liquid-filled suit/intersuit estimated mission duration.

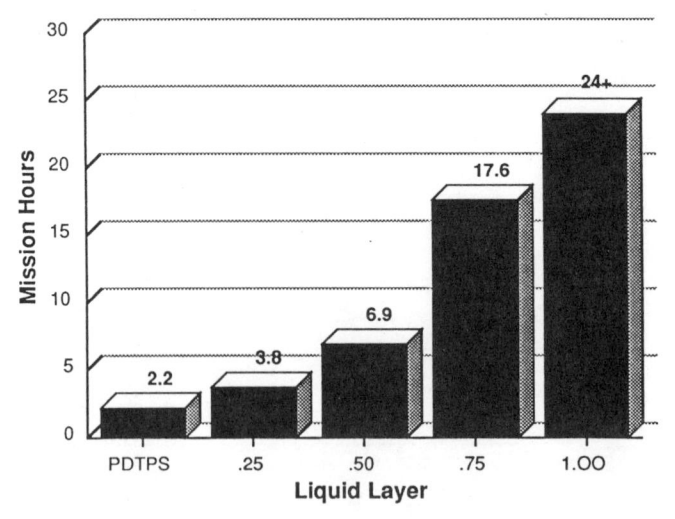

Figure 4. Liquid-filled suit/intersuit estimated mission in 28°F water.

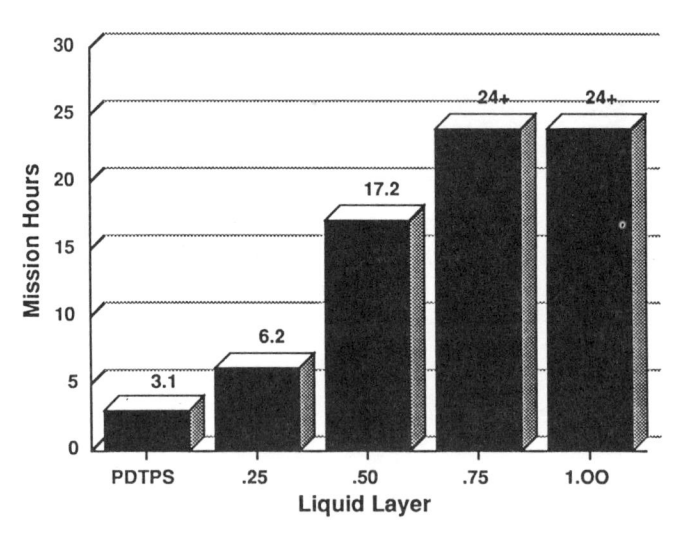

Figure 5. Liquid-filled suit/intersuit estimated mission in 35°F water.

Section 3

Chemical Hazards

Protection of Divers in Waters Which Are Contaminated with Chemicals or Pathogens

J. E. AMSON
Office of Marine and Estuarine Protection
United States Environmental Protection Agency
Washington, D.C.

Introduction

Toxic and anthropogenic wastes are present on the coasts and in the coastal waters of the United States, particularly along the East Coast. Medical and drug-related wastes, such as syringes, crack vials, and hospital dressings washed ashore during the summer and early fall of 1988 from Florida to Massachusetts and resulted in numerous beach closings for health and safety reasons. Over 800 bottlenose porpoises died in the coastal waters of the eastern United States during the summer of 1987, most of which showed severe necrotic epidermal lesions. There are abundant accumulations of plastic trash on the beaches of the United States, including tampon applicators, six-pack beverage rings and Styrofoam drink containers. A "brown tide" which severely damaged the bay scallop industry and the eel grass population occurred in Peconic Bay, New York, during the summers of 1986, 1987, and 1988. A "green tide" which closed the beaches and caused substantial discomfort to the community occurred off Ocean City, NJ, during the summers of 1985, 1986, and 1987. Over one-third of the nation's shellfish beds are permanently or intermittently closed due to bacterial contamination. Numerous beaches on the east coast of the United States are closed during the summer months due to unplanned discharges or storm sewer discharges of potentially pathogenic sewage. Finally, lobsters caught for human consumption off the northeastern United States had to be discarded due to the necrotic tissues.

These events have caused great concern among public and government officials about the health of our coastal waters. Contamination may arise from a variety of sources, such as discharges from industry or municipal outfalls, runoff from nonpoint sources, or spills of hazardous material. Such materials as plastic precursors, synthetic organic chemicals, and pesticides are a common and essential entity of our daily commerce. Many are transported by waterborne tankers, highway tank trucks, and railroad tank cars. Newspaper and television accounts suggest that accidents involving these varied forms of transport are a common occurrence.

Exposure of Divers to Contaminated Environments

During the last decade, diving operations in polluted water have increased the range of materials to which divers have been exposed. At first, little consideration was given to the possible effects of such materials on the divers themselves. For example, a number of years ago the National Oceanic and Atmospheric Administration Diving Office was asked to review a research proposal to put divers down to 60 feet in the New York Bight, and then dump sewage sludge on them from a disposal barge to film the dumping action from underwater.

In the late 1970s, the attitude toward exposure of divers to pathogens and chemicals began to change after NOAA completed a study which showed that pathogenic microorganisms, such as bacteria, viruses, and parasites, clearly posed potential hazards for divers in ocean-dumping areas. The results were corroborated by several incidents, including one in July 1982 when a number of New York City firefighters and police officers contracted amoebiasis after participating in diver training exercises in the Hudson River near a discharge pipe for raw sewage.

Today, divers are occasionally required to enter contaminated waters to assess damage from leaking vessels or pipelines; to locate, contain, or clean up underwater sources of contamination; to recover drums of potentially hazardous substances; or to conduct research studies. Submersion in these situations has resulted in injuries such as chemical burns, both to divers and surface support personnel handling contaminated equipment, as well as in serious illness in divers required to inspect or repair sewage outfall pipes. For example, exposure of rubber-based diving equipment to environments which were contaminated with petroleum products has resulted in diver injury due to equipment deterioration by the petroleum products, and subsequent failure of the equipment.

Clearly, identification of the potential hazards a diver may encounter is critical prior to the start of any diving effort (1). The range of hazardous materials is extensive, but most can be categorized into one of four groups: petrochemicals, chlorinated hydrocarbons and related halogenated organic compounds, noxious gases, and strong acids and bases. The toxicity of these materials varies markedly; exposure to halogenated organics in the aquatic environment, for example, can be extremely dangerous, since some of these compounds can penetrate both suit materials and the diver's skin.

The identification of the hazardous material in an aquatic discharge may be determined in several ways: from a shipping manifest, by contacting the carrier, or by other direct means. If positive identification cannot be made, sampling and analysis are required. Sampling the chemical hazard directly from its container is preferred. If direct sampling is not possible, samples should be taken directly downstream of the source of the hazard. Similarly, microbiological surveys of areas where diving operations are proposed must be undertaken to provide information on potentially pathogenic organisms, and their virulence. It should be borne in mind that it often takes several days to isolate, culture, count, and type the microorganisms; the information will not be available immediately.

There are a number of frequently transported chemicals that are so potentially hazardous that no exposure should be considered. These chemicals have very high dermal penetration rates, and are systemic or central nervous system poisons. Table 1 lists chemicals in the aquatic environment to which a diver should never be exposed and chemicals to which exposure should be absolutely minimal in such extreme situations as to preserve human life, or to prevent massive environmental damage. The two lists in Table 1 are by no means complete, and any chemical encountered in a field situation should be evaluated on an individual basis with the Environmental Protection Agency Oil and Hazardous Materials Technical Assistance Data System. The EPA Oil and Hazardous Materials Technical Assistance Data System is the responsibility of the Emergency Response Division (WH-548B), Office of Emergency and Remedial Response, U.S. Environmental Protection Agency, Washington, D.C. 20460, which lists, among other parameters, material solubility in water, material corrosivity to various substances, and allowable exposure concentrations.

Equipment Considerations

There are two basic methods of diving: One-atmosphere diving and ambient diving. In one-atmosphere diving, the diver is enclosed in a rigid suit that contains air at the surface pressure of 14.7 psi. Recent developments in one-atmosphere suits have concentrated on articulated metal suits, which allow divers to work in a one-atmosphere environment at depths exceeding 1000 feet. However, such suits are simply too bulky and unmaneuverable for most underwater efforts.

Table 1.

Chemicals to which divers should never be exposed
- Acetic anhydride
- Acrylonitrile
- Bromine
- Chlordane
- Epichlorohydrin
- Methyl parathion

Chemicals to which exposure should be absolutely minimal
- Benzene
- Carbon tetrachloride
- Cresol
- Dichloropropane
- Ethyl benzene
- Hydrogen sulfide
- Methylene chloride
- Methylmethacrylate
- Naphthalene
- Perchloroethylene
- Polychlorinated biphenyls
- Styrene
- Toluene
- Triochloroethylene
- Xylene

In ambient diving, the diver is subjected to the ambient pressure of the water depth to which he has descended. Ambient diving is divided into two subgroups: self-contained diving and surface-supplied diving. Self-contained diving utilizes scuba, and is widely used in scientific research. The diver's air supply is provided from a regulator held between his teeth so the diver's mouth is constantly exposed to the water. In addition, the inhalation of contaminated microscopic water droplets from the exhaust valve of a scuba regulator provides a direct passage of the contaminant to the lungs, and thus to the bloodstream. Further, the only way for a scuba diver to clear condensation from inside his mask is to flood it with surrounding water, thus exposing the nose and eyes to the water. The limitations of scuba in contaminated waters for ocular, nasal, and oral exposure are obvious. The standard scuba dive systems are simply inadequate to protect divers in contaminated waters.

Surface-supplied diving consists of a rigid helmet attached to a waterproof suit. There are many variations of helmets and suits on the market today. Surface-supplied diving provides a major advantage over self-contained diving: the diver's air supply is provided through an umbilical hose, and thus, the diver is not limited by the amount of air carried in a tank on his back. Since there must be an umbilical to supply the breathing gas, surface-supplied diving has the additional advantage that a heating (or cooling) water hose that feeds the diver's suit can be in parallel with the air hose. In addition, a hard-wire link for constant communication between the diver and his surface tenders and a safety lifeline are married to the umbilical for retrieval of the diver in case of an emergency.

A diver's protective suit should have strength, flexibility, ease of decontamination, and most important, chemical resistance. The material from which a suit is constructed will have a considerable effect on the amount of pollutant that will be absorbed or passed through the suit. Several types of protective suits are available, ranging from foam neoprene rubber, such as a standard scuba wet suit, to smooth neoprene rubber suits, such as the Draeger suit, to suits made by the Viking Rubber Corporation, to the recently developed Suit-Under-Suit.

Foam neoprene is a poor choice for diving in chemically contaminated environments because it can act as a sponge and absorb large amounts of contaminated water. In addition, certain contaminants can degrade foam neoprene and pass through to the diver. This is particularly true during decontamination procedures, when high pressure sprayers, used in the initial cleaning of the diver and the suit, may force the contaminant through the foam neoprene. Smooth neoprene suits, such as the Draeger suit, do not absorb contaminants and are more easily decontaminated than foam neoprene suits. However, the suits are fairly thin, provide no thermal protection, and have a tendency to tear or puncture. Suits made by the Viking Rubber Corporation are made of very heavyweight rubber bonded to a polyester fabric. The suit is fully waterproof but, like the Draeger suit, provides no thermal insulation. The diver must wear insulated underwear for warmth and protection against chafing from the inner fabric of the suit. In warm aquatic environments, or in the case of maximum exertion by the diver, the Viking suit can be a substantial disadvantage because there is no way to cool the diver inside the suit. The Viking suit has a major advantage over the Draeger suit because it is very resistant to tearing; in addition, it is more easily decontaminated than a foam neoprene suit.

The Suit-Under-Suit (SUS) dive system, developed jointly by EPA and NOAA, provides substantially improved protection for divers. This positive-pressure dive suit provides an innovative solution to two problems associated with contaminated water diving: thermoregulation and leakage. The SUS consists of an inner, thin-rubber dry suit with attached boots. A second, looser, modified Viking Suit with exhaust valves on both ankles and on one arm is worn over the inner suit. A neck dam on the outer suit is mated to a similar device on the inner suit, thereby creating a closed cavity between the two suits and separating the diver's head from the two suits. Clean water of the appropriate temperature is pumped through one of the umbilical hoses into the cavity between the two suits to warm or cool the diver. The water exits through the ankle and arm exhaust valves in the outer suit. Since the cavity between the suits is filled with water under a pressure slightly greater than that of the ambient water, any puncture or leak in the outer suit results in clean water leaking out, rather than outside water leaking in, as would be the case in air-filled suits such as the Draeger or nonmodified Viking Suits.

Finally, it should not be assumed that because a diver has received initial training in a dive system he is proficient in the use of that system. Continuing retraining is necessary if the diver is to remain proficient in the use of the system under all circumstances.

Decontamination Procedures

After every dive into contaminated or potentially contaminated water, the diver must be decontaminated to avoid danger to himself or to the personnel handling the equipment. The first step in the decontamination process is to wash the diver down with a high-pressure spray to remove any adhering contaminants or residues; the second is a washdown with a surfactant, such as trisodium phosphate, or with a solvent appropriate to the contaminant to which the diver has been exposed. If the diver has been exposed to water contaminated with pathogens, the second step should be followed by spraying down the diver with a clinical disinfectant such as Betadine Surgical Scrub Solution. Finally, the diver should be sprayed with fresh water to remove the final decontaminant. Areas that need special attention during decontamination procedures include zippers, seams at junctions of suit surfaces, helmet-sealing mechanisms, and the soles of the diver's boots.

In decontaminating a diver, the preferable washdown pattern is top-to-bottom with the sprayer nozzle facing downward; it is also important not to touch the diver with the sprayer nozzle to prevent contamination of the nozzle. An optimal distance of the sprayer from the diver is 1.5 to 2.0 feet; this distance also reduces the splashback of contaminants that may hit personnel in the decontamination area. Decontamination personnel must also be protected from the contaminants they are washing from the diver. In addition, commercial recovery operations have shown that recovery of sunken chemical drums and containers by divers can often lead to contamination of the dock or the ship's deck and surrounding equipment. Care must be exercised to ensure that surface personnel do not spread the contaminants beyond the immediate decontamination area. Finally, it should be evident that solutions resulting from diver decontamination should not be routinely discarded; such solutions should be treated as diluted portions of the hazardous contaminant, and treated accordingly.

Revised EPA Diving Safety Policy

One development that has led to increased diver protection, in safe as well as in contaminated and polluted waters, is the recent revision of EPA's Diving Safety Policy (DSP). This revised Policy is consistent with the regulations of the Occupational Safety and Health Administration (OSHA) and with NOAA's diving requirements. (The OSHA Regulations for commercial diving operations are found at 20 CFR 1910, Subpart T). (NOAA's diving requirements are found in NOAA Directive 64-23, dated November 30, 1983, issued by the Assistant Administrator for Ocean Services and Coastal Zone Management, National Oceanic and Atmospheric Administration, Rockville, MD 20852. The revised DSP is officially chapter 10 of EPA's Occupational Health and Safety Manual of the Environmental Health and Safety Division (PM-273), Office of Administration, U.S. Environmental Protection Agency, Washington, DC 20460. The new policy accomplishes several main objectives: (a) It establishes EPA policy for all diving operations, in accordance with the OSHA regulations. (b) It ensures that all diving operations performed by EPA, or its contractors, are conducted in a safe manner, with uniform procedures, and by sufficiently trained personnel. (c) It applies to all diving operations carried out by any employee of EPA during the course of official duties. The DSP also applies to any non-EPA employee engaged in a diving operation under the auspices of EPA. (d) The DSP establishes an EPA Diving Safety Board (DSB), composed of all Unit Diving Officers.

The DSB is responsible for, in part: (a) establishing policy and operating procedures to ensure a safe and efficient diving program; (b) reviewing existing policies, procedures, and needs to ensure a continually high level of technical skills and knowledge; (c) establishing policy for the initial certification of divers and refresher training of experienced divers; and (d) reviewing EPA diving accidents or potentially dangerous situations, and establishing preventive measures to ensure the avoidance or recurrence of such incidents. The Policy establishes qualifications and responsibilities for the Board Chairperson, a Board Technical Director, a Board Training Director, as well as Unit Diving Officers, Divemasters, individual divers, and dive tenders.

Conclusion

Through the work that has been done in recent years by EPA, NOAA, and other agencies, contaminated and polluted water diving has evolved into a distinct form of specialized diving. This evolution is continuing, and more developments can be expected in the future. Increased demands will be put on divers and diving systems by continuing efforts to clean up waters that have been receptacles for chemical, toxic, or pathogenic wastes.

The safety of the diver and support personnel must be the primary concern of any clean-up or emergency response effort. Personnel should be made aware that standard scuba dress offers minimal or no protection in many contaminated waters. Every plan for a dive into waters which contain hazardous or contaminating materials must be carefully evaluated and weighed against the short-term and long-term hazards of the particular contaminant involved.

Acknowledgment

The author is indebted to Rick Traver, of EPA's Hazardous Waste Engineering Research Laboratory, in Edison, NJ.

References

1. Traver, R.P. 1968. "Interim Protocol for Diving Operations in Contaminated Water," EPA Report 600/S2-85/130, Hazardous Waste Engineering Research Laboratory, Cincinnati, OH 45268.

Notice: Mention of commercially available products in this paper does not constitute certification or approval of these products by the United States Environmental Protection Agency or the United States Government. The views expressed in this paper are those of the author, and not necessarily those of the U.S. Environmental Protection Agency.

Section 4

Equipment and Procedures

Practical Systems for Contaminated Water Diving

STEVEN M. BARSKY
Marine Marketing and Consulting
Santa Barbara, California

Introduction

During the past few years, standard procedures and protocols have been well defined for diving in contaminated water under a variety of conditions (1) such as biologically, chemically, and radioactively contaminated environments. Divers working in these environments have represented fire departments, law enforcement agencies, academic institutions, commercial diving firms, and the military. Unfortunately, there are still many cases of divers working in contaminated environments without proper protection. Incidents continue to occur for a variety of reasons, including lack of education, "machismo" (when divers themselves underrate the hazards involved), lack of funding for proper equipment, or when there are concealed hazards, either because of deliberate concealment or lack of awareness by those responsible for the hazards.

Equipment which is appropriate for a particular dive by one dive team may not work at all for another team but when selecting a system for use in contaminated water diving, certain minimum technical specifications must be met. Conversely, while a system may be technically excellent, it may be impractical because of expense, logistical considerations, or training requirements for a particular group. Further, if a system is uncomfortable and fails to gain diver acceptance, its potential worth is nil, regardless of its capabilities.

Diving System Selection

Optimum life support is "not to enter the water" (2). However, contaminated-water diving is almost always necessitated by some type of emergency, so there is rarely a choice on whether or not to undertake a dive except when injury or death is likely.

The problem in selecting a system for a particular situation depends on defining "contamination." Diving in the open harbor of a small coastal town in New England is significantly different from diving in the New River, which borders Imperial County (San Diego) and Tijuana, Mexico, where raw sewage and agricultural chemicals are prevalent. The questions which need to be asked are "What works?" and "What's available?"

For manned diving, equipment can be assigned to one of four general classifications.

1. Advanced Diving Systems and Submersibles. The best protection in contaminated water is offered by submersibles and advanced diving systems such as JIM or the Newt Suit. Unfortunately, the financial costs for purchase, handling, maintenance, and training of personnel in the use of such systems puts them beyond the range of any except the largest government agencies and commercial diving firms.

2. Surface-Supplied Positive-Pressure Systems. Systems such as the Navy MK 12 offer the highest degree of protection for ambient-pressure manned diving. The advantage of an unlimited air supply with a positive-pressure capability cannot be denied. Although the helmets are always under positive pressure, the lower portions of the diver's suit see an underpressure, a problem if the suit is punctured unless the diver is equipped with the Suit-Under-Suit (SUS) system as developed by NOAA and the EPA (1). At this time, the SUS system is not commercially available, and most municipal diving operations do not have the technical exper-

tise to rig such systems or the logistical support required to operate them. The SUS system requires 7 liters per minute of clean water to maintain pressure in the suit. Additionally, donning the SUS suit takes much longer than other systems and may impose a heavy work and thermal load before the diver even enters the water.

Positive-pressure systems have a number of liabilities which make them impractical for many operations. First, the high-volume, low-pressure, diesel-driven compressors, which are typically used for surface-supplied diving, require handling by forklift or crane and preclude transportation by automobile or inflatable boats. Second, the cost is usually much higher for comparable surface-supplied demand systems. Additionally, the training required with these systems is much more complex than for scuba or surface-supplied systems. The principal reason for the increased complexity is that the the diver's life-support system and buoyancy are both regulated with the same controls (non-SUS mode). Divers must provide themselves with sufficient breathing air without becoming overly buoyant. At the same time, the diver must be attuned to subtle changes in the volume of air passing through the helmet as the compressor cycles off and on in response to the pressure changes in the breathing supply. Inattentive divers may suddenly find themselves positively buoyant and heading for the surface.

3. Scuba Systems for Contaminated Water. Many public service agencies use scuba systems for contaminated-water diving. At a minimum, these systems usually include a positive-pressure full face mask, a dry suit specific to the hazard level, dry gloves or mittens, and some type of tethered communications or wireless system. Such a system is used by the New York City police department for diving in the Hudson River on a daily basis and was used when the Rhine River in Germany was plagued by a serious toxic spill several years ago.

Full face mask scuba systems are often selected for reasons of portability, simplicity, and economics. However, although they

Figure 1. Positive pressure suit and Navy MK 12 helmet.

Figure 2. Scuba system for contaminated water.

may be acceptable for some situations, full face masks are not the best defense in serious contamination. If a full face mask is used, it should incorporate a positive-pressure regulator which will free flow and expel any water should the seal between the face and the mask be broken. Even with the best full face mask, there is danger of the mask coming off.

4. Surface-Supplied Demand Systems. These are probably the most practical for the public safety diver or any agency which demands portability. The minimum components include a demand-diving helmet equipped with some type of double exhaust system (two exhaust valves mounted in series); a hazard-specific dry suit equipped with dry gloves or mittens; a compact dive console incorporating a pressure regulator, communications and depth sensing; a high-pressure air supply; and a diving umbilical.

The standby diver must be equipped with gear at least equal to that worn by the diver in the water. It is unacceptable for the standby diver to be equipped with only conventional scuba, but some public safety agencies persist in this practice due to economics or ignorance.

The advantages of a surface-supplied demand-diving system include portability, ability to operate with a limited air supply, minimum-system deck-space requirements, relatively low cost, and minimum training requirements. Trade-offs with surface-supplied demand systems center around the exhaust valves on both the helmet and the suit, which are always a potential ingress point for contaminants.

Guidelines for Dry Suit Selection

No matter which diving mode is utilized, certain guidelines should be followed in selecting a diving suit for a contaminated-water dive. Some considerations are specific to certain types of contaminants, while others are common to all situations. It is generally recognized that the suit should have a smooth, nonporous outer surface rather than a fabric-coated surface which will trap contaminants. Currently, the most popular type is made from a combination of natural and synthetic rubber, vulcanized into a seamless, one-piece suit which will work well for both biologically and radioactively contaminated environments. However, it will not protect the diver from every type of chemical hazard; the suit must be specific to the chemical hazard.

Other requirements include provision for attachment of dry gloves or mittens, attached boots, and automatic buoyancy/exhaust valves for suits used with demand helmets. Both the suit and spares must be commonly available; good user documentation must accompany the suit and the suit must be easy to decontaminate. Further, the design must be ergonomically sound, permitting the diver freedom of movement to swim, climb a dive ladder, and work without interference.

Guidelines for Demand-Diving Helmet Selection

Demand-diving helmets for polluted water must also be selected with an understanding of the environment in which they will have to operate. To prevent any backflow of contaminants, the helmet must be equipped with a series or double exhaust mechanism. If the helmet is equipped with both a regulator exhaust and a main exhaust, the two must be "tied" together.

From a human engineering standpoint, the helmet should be neutrally buoyant with low internal volume. Communications, as with any surface-supplied diving operation, should be excellent. Clear, concise support documentation should be supplied with the helmet. Spare parts should be commonly available, and the helmet must be easy to decontaminate, particularly important considering the number of valves and penetrators common on most diving helmets. Molded handles on valves and other life-support components often have channels and other cavities which are

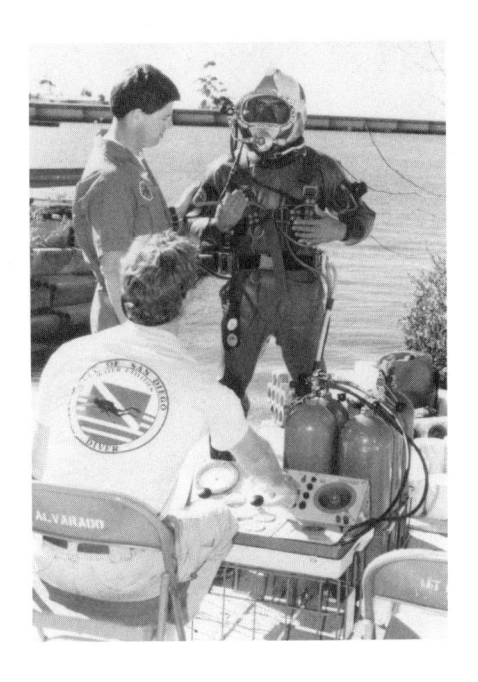

Figure 3. Surface supplied diving system.

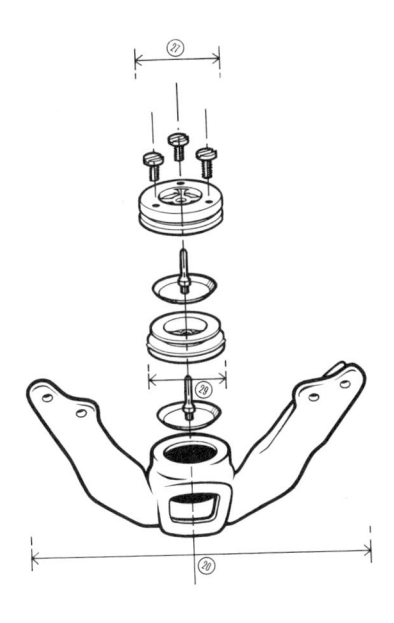

Figure 4. Double exhaust system for demand diving helmet.

153

essential for manufacturing but may trap contaminants. Additionally, helmet "O" ring and diaphragms should be examined to ensure chemical compatibility where needed and inspected prior to every dive.

Helmet/Suit Interface Guidelines

Since no one manufacturer is capable of providing both helmets and suits for contaminated environments, equipment must be selected carefully for optimum interaction. Some systems benefit from cooperation between firms, while other pieces of equipment give evidence of a lack of understanding of the way in which the components must work together. Obviously, any helmet/suit interface must be watertight, but this is not achieved easily. Furthermore, the interface should not be allowed to trap any contaminants. Tenders must scrupulously examine the joint where the helmet attaches to the suit: the diver, while being undressed, may be exposed to contaminants trapped in this junction.

High reliability is essential for the entire system, but particularly for the interface between helmet and suit. Maintenance requirements should be low, and the system must be capable of operation under demanding conditions. Conversely, it should be recognized that exposure to certain environments will require replacement of the whole system because of structural changes in the equipment from chemical exposure or simple mechanical wear. One-third of all suits and other soft components should be budgeted for annual replacement.

The suit inflation system used in surface-supplied demand suits must be fail-safe. Should the suit inflator hose rupture, no contaminants can be allowed to enter the air supply system. The suit inflation system should be independent of the diver's bailout supply. Isolation of the diver's head and life-support system from his body and suit environment is highly desirable and essential when demand-diving helmets are used. Should the diver's suit be

Figure 5. Dry suit mating yoke for demand diving helmet.

damaged, isolating the head will prevent ingestion of contaminants and provide protection for the eyes and respiratory system. Further, since the dry suit would function as a breathing bag if the helmet and suit communicate directly, a neck dam must separate the two for the demand regulator to function properly.

References

1. Traver, R.P., J.M. Wells. 1984. Summary of on-scene coordinator protocol for contaminated underwater operations. pp. 73-89. In: Proceedings of the Fourteenth Annual International Diving Symposium. 73-89. Association of Diving Contractors, New Orleans.

2. O'Neill, W.J. 1974. Safety consideration in undersea life support. pp. 341-351. In: Proceedings of the Working Diver. Marine Technology Society, Washington, DC.

3. Coolbaugh, J.C., O.P. Daily, S.W. Joseph, R.R. Colwell. 1981. Bacterial contamination of divers during training exercises in coastal waters. Marine Technol. Society J. 15:15-22.

Diving in Nuclear Power Plants

RANDY THOMPSON
Solus Schall
1441 Park 10 Boulevard
Houston, Texas

Introduction

Oceaneering International, Inc., together with its affiliate companies, is the world's largest publicly owned underwater services company which specializes in manned and robotic underwater services, survey and positioning, search and recovery, engineering, project management, and inspection. The majority of services are provided to the offshore oil and gas industry and include underwater drilling support; subsea engineering and construction; production systems management; and facilities inspection, maintenance, and repair.

Underwater maintenance in nuclear power stations is becoming an established and very sophisticated segment of the nuclear service market. There are approximately 105 nuclear power plants in commercial operation in the United States, which produce over 17.7% of the electricity used in the United States, and there are 11 reactors with construction permits. Reactors use uranium fuel assemblies to produce heat, which in turn boils water; the steam turns turbines to produce electricity. The fuel assemblies are stored and re-fueled underwater in the fuel storage pool, which contains very clear water approximately 40 feet deep and has a lining constructed of stainless steel, the metal used for the majority of the equipment in the pool. Filters are used to remove radioactive contamination and maintain excellent visibility.

Divers are used to remove and replace fuel racks, inspect and repair transfer equipment, and perform inspections of equipment and structures in the pool. The two most important concerns of the diver are radiation and temperature. Radioactivity, the emission of energy from the nucleus of an atom, has 4 types: neutron, beta, alpha, and gamma. Beta and gamma radiation concern the diver. Neutrons are only a factor while the reactor is operating, and alpha radiation only travels about 1 inch in air and can be stopped by a sheet of paper.

Radiation dose is usually measured in units of rem. Repairs to fuel-transfer systems, fuel-handling bridges and spent-fuel pool modifications usually result in less than 1 man-rem exposure per event. Longer re-rack operations usually result in from 2 to 20 man-rem units for dive teams working in the presence of spent-fuel assemblies for periods extending to several months. However, from the standpoint of ALARA (as low as reasonably achievable) radiation exposure, man-rem exposures for divers are about 40 times lower than when doing the same job topside in the dry.

Exposures are limited by reducing the time spent near a source, increasing the distance from the source, and placing shielding between the source and the worker. The exposure to personnel is measured by personal dosimetry devices such as the self-reading pocket dosimeter, the TLD badge or a film badge. The federal government has standards for the amount of radiation an individual is allowed to receive in a given period of time. In some storage pools, the water temperature is over $10°F$, and since the diver is wearing a dry suit, there a possibility of dangerous hyperthermia. We utilize a cool vest worn under the dry suit to circulate chilled water to the diver's torso.

Guidelines

Oceaneering International has guidelines for radiation protection practices for diving in contaminated water or near sources of radioactivity. The limits and precautions are: (a) A valid Radiation Work Permit (RWP) should be approved before any diving operation is allowed. The RWP should specify all protective clothing requirements and diving dosimetry as required. (b) No diver is allowed to come within 20% of his current quarterly exposure limit without special permission from the station health physicist. (c) A body burden analysis (BBA) is performed as required by the Health Physics Manual, and a "baseline" urine sample is taken and recorded when a diver first arrives on site (before diving), and another sample is taken before he leaves the site after completion of the diving project. (d) All diving personnel must have an in date diving medical on file and recorded in their dive logs. (e) Individual radiation records from previous exposures are maintained in each employee's permanent personnel file. (f) Training in health physics and security for each job is governed by individual client requirements. (g) All personnel are familiar with all local regulations covering exposure to radiation: (10 CFR 19, Workers Rights and Responsibilities and 10 CFR 20, Exposure Limits apply to all United States Nuclear Operations).

Due to radioactive contamination in the water, special equipment is required: (a) In some cases, divers cannot exhaust bubbles which would release airborne contaminants into uncontaminated areas. Therefore, Oceaneering International can apply one of its existing technologies to solve the nuclear plant's problems, the gas reclaim hat to the surface; the "Rat Hat" diving helmet (nuclear modification) allows the diver's exhaust to return to the topside manifold. The Viking Dry Suit (nuclear, special order) has been fitted with the Oceaneering Rat Hat dry suit collar. No inflation or exhaust valves are installed on the suit. The inside wrist sealing ring is permanently mounted to the cuff. Protective gear is installed to prevent chafing. (b) Other special equipment are a large standard diving harness, a weight belt with optional ankle weights, and at least two pairs of rubber gloves, which are secured with electrical tape to the suit cuffs. (c) The diving hose is married to a mil-spec communication wire with electrical ty-wraps. Eliminating duct tape and the rope strength member helps to reduce the fixed contamination picked up on the hose. A pneumo hose is usually not required, as the depth of water in the nuclear power plant's reactor vessel or spent fuel pools rarely exceeds 40 feet. (d) Main and stand-by supplies of filtered, breathable air is provided by compressor or high pressure air banks via a volume tank and manifold. The use of in-plant air, if of breathable quality, should be investigated. (e) Standard dive radios can be used, but a smaller portable belt-mounted radio is recommended. (f) A safe means of entering and leaving the water is necessary. Persons entering a radiation area or contaminated area are required to wear anti-contamination clothing. Divers entering the fuel storage pools are required to wear various combinations of dosimeters and alarms. Divers should wear personnel dosimetry on the head (inside/outside the helmet), chest, back, gonads, right thigh, left thigh, right ankle, left ankle, right hand, left hand, and right arm, left arm (above the elbows). The dosimetry packages should not include the diver's normal monthly TLD or pocket dosimeter.

Procedures

Dressing in and dressing out procedures must be written, a radiation survey of the pool must be done, and the suit and helmet must be checked for leaks. Dressing in the diver can take up to an hour.

Diving operations should be planned with appropriate groups prior to beginning work. Instruments in the general area should be surveyed, including an underwater radiation survey.

The plant air should be sampled after filtration (to ensure that it meets quality standards for breathing air) before using it for breathing in diving operations. In meetings with divers, it is important to ensure that they have read and understand the requirements of the procedure prior to their first dive. and to discuss the safety, surveillance requirements, contamination control measures to be used, the scope and duration of the planned work during the dive, and the results of the surveys. It is important to be sure that the divers dress according to the work plan.

The high pressure technician should have a telescopic dosimeter to monitor the diver when he enters the water. If the water clarity is bad, the work area is not accessible for survey from the surface, there is floating material in the water, or radiation levels are fluctuating, the diver should carry an underwater survey instrument or a detector that is capable of being read out by the health physics technician at poolside or by the diver. When the diver has completed his task he is washed down with demineralized water, monitored for contamination, and dried off with clean absorbent material before he can remove his diving equipment. Remove the head dosimetry device from the diver's helmet (inside and outside) as he removes his diving equipment. When the diving operation is finished, diving equipment is surveyed, decontaminated, tagged, wrapped, discarded, or stored. The final process is to analyse body burden and urine.

Example

All usual surface jobs can be tackled by underwater operations: cutting, welding, non-destructive testing, visual inspection, assembly or removal of mechanical structures, equipment setting or relocation, cleaning, decontamination, to name but a few. The major areas in which divers have been used include spent fuel pool re-racking, various inspection and repair operations on fuel transfer systems, internals, spent-fuel pool, fuel-handling equipment, also leak detection on storage tanks, and retrofit of fuel handling systems.

An example of a diving job would be to increase the capacity of the spent-fuel storage for a nuclear power station by the removal of the old contaminated low-density racks and installation of new high-density racks. A recent re-rack job consisted of removing 502 individual cells and supporting grid work and installing 9 free-standing high-density racks approximately 15 to 20 feet high. This was approximately a three-fold increase in the fuel storage capacity. The job took 7 and a half months and required 224 dives by five divers, an average bottom time of 2 1/4 hours, and an average whole-body gamma dose of 11.12 mRem per dive or 5 mRems per hour of bottom time. This particular re-rack operation utilized Plasma Arc Cutting (PAC) in place of conventional oxygen-arc cutting, thus reducing chemical pollution and clarity problems, and increasing cutting efficiency 10 fold. Approximately 1200 linear feet of cells and grids were cut, saving approximately $250,000 in excess vacuum filters for the slag residue. Involvement during the design phases of the project enabled us to develop a lift bag which saved the power company several hundred thousand dollars in alterations and additional man hours as well as reducing safety problems by eliminating the need to remove the fuel-handling bridge to access the end of the spent-fuel pool.

Conclusion

With proper planning, procedures, training, and equipment, divers can perform many tasks in the nuclear industry safely and efficiently at considerable savings. Similar operational procedures, personnel and equipment can cross over into the problems of diving in environments that involve biological and chemical hazards.

Safe and Cost-Effective Methods of Diving in Contaminated Water in the Midwest

MIKE MCGOVERN AND GREG ROBERTS
Midwest Marine Contracting, Inc.
Kansas City, Missouri

Introduction

Typical missions for a diver in the Midwest occur where the water can harbor a variety of submerged and floating substances, including petroleum products such as fuel and creosote, solvents, sewage, PCBs, hazardous chemicals, and many unknown substances. Until recently, so little was known about diving in contaminated environments that appropriate protocols were not developed. Since divers lacked knowledge and awareness of contaminants and their effects, they would ignore the hazard unless they developed problems after the exposure. Aside from being required to keep their tetanus immunization current, divers in contaminated environments generally did not deviate from standard equipment and procedures. Despite a growing number of requests for diving operations where there are known contaminants, water conditions continue to be classified by clarity rather than by possible contaminants, which has prompted many contractors to attempt to determine precautionary measures and protocols for diving in contaminated water.

Classification

Diving in contaminated water has been taking place using standard procedures and equipment for many years. Typical sites that involve contaminated water include wastewater and nuclear power plant facilities, refineries, paper mills, spills of hazardous materials and petroleum products, construction sites, and water treatment facilities. There are three levels of water contamination: Level I is characterized by no known contaminants; an operation in Level I water calls for standard surface air supplied (S.A.S.) with a mask, a wet suit, coveralls, boots, and gloves, as well as standard cleaning and maintenance of equipment. Level II is characterized by intermittent or continuous exposure to contaminants harmful to the skin, for example, placing concrete or epoxies, or being exposed to certain petroleum products, which include dives at sewage treatment plants or nuclear plants; it requires S.A.S. with a dry suit, a helmet, coveralls, boots, gloves and protection of tenders. Level III is characterized by known health hazards that are serious and requires avoiding exposure of the skin or ingestion of the contaminated water, such as dives at sewage treatment plants or nuclear plants; these operations warrant S.A.S. with a Viking-type suit, an attached helmet with series exhaust valve, sealed gloves, coveralls, and boots and tenders should also wear protective clothing. Decontamination procedures are mandatory for Level III operations.

To be able to select the appropriate equipment, personnel, and procedure for an operation, the dive supervisor must determine the degree of exposure and the contaminants involved. If this is not possible prior to starting the operation, the supervisor should assume and prepare for the worst, for even then, conditions may not permit a safe dive. Surveying the site, gathering samples, and consulting with experts is the only way to conduct an operation safely. Once the supervisor has determined that conditions are conducive to a safe operation, he or she must set a procedure for all to follow.

Equipment Selection

Diving in contaminated water at Level I or Level II requires nothing more than off-the-shelf equipment and commonsense procedures. The Level III classification, however, calls for maximum protection of divers and tenders throughout the operation. To choose suitable equipment, the supervisor should (a) locate diving and industrial suppliers with readily available supplies at reasonable cost, (b) use helmets with such features as protection, visibility, adequate ventilation, easy maintenance, proper seals, and good communications, (c) use dive suits with such features as thermal regulation, positive seals, attached boots, attachable gloves, and a minimum of seals, (d) ensure chemical compatibility (retention and permeability), (e) establish simple dressing and undressing procedures, and (f) enforce simple maintenance and disinfecting and decontamination procedures.

When diving in Level III water, standard equipment must be modified to increase protection and performance. The following are examples of such added protection: A free-flow helmet enables positive pressure to be applied at all times through a regulating exhaust valve, installation of a series exhaust valve eliminates splashback, four-wire communication allows continuous monitoring of the diver, a positive clamping device and a secondary glove system helps to seal gloves properly, a light layer of protective clothing with a hood under the dry suit helps divers avoid contamination when they unsuit and keep the suit from chafing them, an outer garment and boots protects the suit from excessive wear and puncture, especially in the cuff and zipper areas.

Depending on conditions, an operation may require more modifications than those listed. In any case, familiarity with the equipment and its proper use will ensure a safer operation. Finally, dive supervisors should keep abreast of improvements in equipment. For example, tests of chemical compatibility of diving equipment with hazardous substances are currently being conducted, and recent reports state that a two-layer suit provides better protection and diminishes risk of hypothermia.

Training and Personnel

Unless the divers have had previous training in another field, the average diver may know very little about hazardous substances, how to handle them, or how to guard against contamination. Divers currently are not required by diving schools or by the Navy to learn proper procedures for diving in contaminated waters. On the other hand, a little knowledge can be a dangerous thing.

Training is as important as personnel selection. Properly trained personnel are more confident and better equipped to prevent accidents. They are familiar with equipment, procedures, and other personnel prior to actual exposure to contaminated water and have had an opportunity to learn from their mistakes in a controlled environment. A training program should focus on a thorough explanation of contaminants and their properties, effects of exposure, precautions and methods of protection, and emergency procedures. On-the-job training is not acceptable for Level III jobs.

Procedures

Procedures for diving in contaminated water should be developed from the diving contractor's "Safe Practice Manual" with assistance from outside experts, including health and physics professionals, hazardous-waste specialists, experienced consultants, and medical professionals. The manual should state specific requirements for training and operational procedures that promote diving safety in contaminated areas. The manual should address at least the following: (a) The importance of detailing a job profile and conducting a prejob analysis meeting. (b) Identification of hazardous substances and contaminants. (c) Conduct of predive and periodic surveys, with respect to sampling and quantity. (d) Listing of all safety precautions. (e) Guidelines for client personnel. (f) Setting a staging area for dive station and contaminated areas. (g) Establishing tag-out procedures for existing equipment. (h) Conducting predive briefings and discuss special considerations. (i) Procedures for suiting procedures for divers and tenders. (j) Procedures for underwater surveys by divers, fact-finding on present conditions. (k) Procedures for post dive and diver evaluation. (l) Procedures for decontamination and disinfecting. (m) Debriefing of divers. (n) Procedures for securing the dive station and cleaning up. (o) Procedures for emergency and safety operations. (p) Change of conditions/abort dive procedures.

Since no manual, no matter how thorough, can anticipate every contingency, the parties involved in a diving project must apply common sense. If in doubt, the supervisor should discontinue the operation until problems have been resolved. Finally, the diving supervisor must insist on knowing every detail of an operation and should be satisfied with the progress before going further.

Conclusion

Data on diving in contaminated and polluted environments have been collected only in recent years and has led to more questions as work in such environments becomes more common. Diving contractors in the past were most often concerned with the effect of contaminants on equipment, but today the safety of divers and tenders is the paramount concern. There is a continuing need for equipment to be tested and evaluated, for procedures to be standardized, and for diving schools and the Navy to teach introductory courses on diving in contaminated environments.

There is also a need for a central clearinghouse of information on contaminants, their effects, and possible solutions to the hazards posed by such substances. The Association of Diving Contractors (ADC) and its members are in a unique position to spearhead this effort by appointing an inshore diving committee to survey the ADC's own membership and other diving companies for information that could be used to set standards for a section on diving in contaminated environments in the ADC Consensus Standards for Commercial Diving Operations manual.

Section 5

Diving Medicine

Infections Associated with Swimming and Diving

GENEVIEVE LOSONSKY
Center for Vaccine Development
University of Maryland at Baltimore
Baltimore, Maryland

Introduction

In the last 40 or 50 years, man's contact with the water environment has grown, heavily influenced by the development and advancement of underwater breathing apparatus; increased recreational and industrial use of fresh and salt waters has exposed man to new environmental hazards. A notable hazard is waterborne infectious pathogens, including those naturally occurring in the water and those introduced into the environment by industrial effluents and human fecal waste. This communication will focus on what is known about the risks of infectious disease to those who work or recreate on the water, with emphasis, where information is available, on diving.

Waterborne Risks

Risks associated with waterborne infections relate directly to the microbiology of the area where exposure takes place, the degree of exposure, and the type of water activity involved (1). Salt-tolerant microbes, which are potentially pathogenic for man, can be found naturally in marine environments (Table 1). These organisms are prevalent in brackish water where the NaCl content is less than 3% along the coastal United States. Some of these organisms, such as the *Vibrio* and *Aeromonas* species, can also be found in freshwater lakes. Pollution has a marked effect on the microbiology of the water (2), adding fecal coliforms, enteric obligate anaerobes, fecal streptococci, coagulase positive and negative staphylococci, and enteric viruses (Table 2). With additional carbon and nitrogen sources, an increase in naturally-occurring organisms may ensue.

One factor affecting the microbial content of water unrelated to pollution is water temperature. Total coliforms in the Anacostia River, for example, drop more than 10-fold during the winter months (3). The number of *Aeromonas* species, which require water temperatures above 15°C to be metabolically active, drops precipitously in winter. Another factor affecting microbial content is the area of the water table. Sediments have 10- to 100-fold higher concentration of bacteria and enteric viruses than surface water.

Table 1.

Naturally Occurring Water Organisms

Achromobacter species

Acinetobacter iwoffi

Aeromonas species

Alteromonas species

Chromobacterium violaceum

Deleya venustus

Erysipelothrix rhusopatiae

Flavobacterium species

Mycobacterium marinum

Pseudomonas marinum

Vibrio species

Table 2.

Pollution-Derived Water Organisms
- *Actinomyces* species
- *Alcaligenes fecalis*
- *Bacillus* species
- *Citrobacter* species
- *Clostridium* species
- *Corynebacterium*
- *Edwardsiella tarda*
- *Enterobacter aerogenes*
- *Escherichia coli*
- *Klebsiella pneumonia*
- *Legionella pneumophila*
- *Micrococcus* species
- *Neisseria catarrhalis*
- *Nocardia*
- *Pasteurella multocida*
- *Proteus* species
- *Providencia stuartii*
- *Pseudomonas* species
- *Salmonella* species
- *Serratia* species
- *Staphylococci*
- *Streptococcal* species
- Enteroviruses
- Rotaviruses
- Hepatitis A
- Non A - non B hepatitis

The degree of water exposure also determines the risk of infections. Data discussed here evolves from infections acquired by swimmers, since there is little or no comparable data published on divers. In a study evaluating an outbreak of Shigellosis at a public beach in Ontario in 1987 (4), the attack rate of disease coincided with the degree of water exposure (Table 3). Those people who did not swim were unaffected; those who had the most exposure (i.e., swallowing water) had the highest rate of disease. A study conducted in Canada during a windsurfing championship (5) found that infections recalled by 79 contestants directly correlated with the number of falls in the water (Table 4). Symptoms included diarrhea in 16%, wound infections in 22%, otitis media in 8%, and conjunctivitis in 18%. Unfortunately, no attempt was made to isolate causative agents in this study.

Coupled with the degree of exposure, the degree of pollution also affects the relative risk of infection. Here again, information is available for swimmers, not divers. A study conducted in Canada correlated microbial content with illness (6); the concentration of fecal Gram-negative organisms and staphylococci was associated with gastrointestinal, skin, eye, and respiratory infections (Table 5). Another interesting finding of this study was that the Gram-negative counts of surface water were all in the acceptable range for swimming (i.e., less than 100 counts per ml of water), suggesting that more than one variable may have been at play in producing the illnesses seen, and that "acceptable" levels of pollution may be misleading markers of safety. Here again, no attempt was made to isolate the infectious agents causing illness.

The types of infections one sees from water exposure are typically occur in areas of the body that are exposed, namely mucosal sites, ears, skin, respiratory organs, and the gastrointestinal tract. The incidence of infections in swimmers and divers in nonpolluted waters is not known, but infections with marine and freshwater organisms have been described (Table 6). Based on composite reviews of case reports in the literature, wound and ear infections are the most common. Wound infections relate to

Table 3. Attack rate of Shigellosis related to the degree of water exposure, adapted from (4).

Exposure	Attack Rate	Percent (%)
Did not swim	0/9	0
Waded	1/6	17
Head under water	3/15	20
Water in mouth	34/55	62

Table 4. Risk of development of infections following polluted water exposure in windsurfers, adapted from (5).

Number of falls in water	Number of subjects with symptoms/total No.(%)	Relative risk
0 (land controls)	8/41 (20)	1.0
1-10	15/34 (44)	2.3
11-20	9/20 (45)	3.5
20-30	10/14 (71)	3.7
>30	10/10 (100)	5.1

Table 5. Relationship between types of illnesses among swimmers in 10 fresh water beaches in Ontario, taken from (6).

Indicator Organisms	Types of Illness	P Value
Fecal streptococci (water)	all	0.016
P. aeruginosa (sediment)	all	0.36
Total staphylococci (water)	skin	0.044
Total staphylococci (water)	eye	0.002
Total staphylococci (water)	all	<0.001
Fecal streptococci (water)	gastrointestinal	0.069
Fecal coliforms	all	<0.001

Table 6. Suspected organisms in infections associated with nonpolluted water.

Wound Infections - usually secondary to trauma
    Erysipelothrix rhusopathiae
    Vibrio vulnificus
    Vibrio parahemolyticus
    Vibrio alginolyticus
    Vibrio damsela
    Aeromonas hydrophila
    Aeromonas sobria
    Mycobacterium marinum
    Chromobacterium violaceum

Gastroenteritis
    Vibrio cholera
    non-01 *Vibrios*
    Aeromonas species
    Pleisiomonas

Respiratory Infections
    Aeromonas hydrophila
    Chromobacterium violaceum

Ear Infections
    Vibrio species

Serious Systemic Infection
    Vibrio vulnificus
    non-01 *Vibrio cholerae*
    Vibrio parahemolyticus
    Aeromonas hydrophila

trauma; serious infections such as bacteremias and osteomyelitis occur rarely, seen, for the most part, in immunologically compromised hosts.

Swimmers in polluted waters commonly develop gastrointestinal and respiratory illnesses. Gastrointestinal illnesses usually occur within 48 hours of exposure and are associated with fecal coliforms and fecal streptococci. There have been rare outbreaks of Shigellosis, Salmonellosis, Yersiniosis, and diarrhea due to campylobacter and the enteric viruses. It must be noted, however, that in many studies etiologic agents have not been pursued. The causes of respiratory illnesses following swimming are not known. Swimmers in polluted water often do have an increased incidence of otitis externa, skin, and wound infections. Although exact incidence is unknown, it is thought that otitis externa is 5 times more common in swimmers than in nonswimmers. Organisms associated with this illness are *Pseudomonas aeruginosa* and *Staphylococci* species. Wound infections usually contain mixed flora.

For divers, even less data are available than for swimmers, and infectious disease morbidity of divers is just beginning to be recognized as a potentially serious problem. Reports of severe infections include a diver in Guam with an open head injury who developed an intracranial infection with *Vibrio aglinolyticus*, a puncture wound on a diver's leg became infected with *Aeromonas hydrophila/sobria*, and *Legionella bozemanii* caused fatal bronchopneumonia in a diver. The infectious disease morbidity associated with diving--the relation between pathogens and sickness--remains unknown. We do not know the relative risks in diving in polluted and nonpolluted waters. We do not know whether the length of a dive increases the risk of infection. We do not know whether long-term exposure in a given geographic area changes the risk of infection.

Conclusion

Some provocative data suggest that divers may be at increased risk for infection (7). Microbial colonization data collected from divers in Maryland's Anacostia River show that *Aeromonas* species are easily cultured from the ear, upper respiratory tract, and skin surfaces of divers after a dive, even if water counts of these organisms are low. From this brief report, it is clear that more needs to be learned about our water environment and risks to divers. By knowing more about incidence of infection, types of infections, importance of variables such as pollution, and degree of exposure we hope to make man's contact with the water safer.

References

1. Auerbachk, P.S. 1987. Natural microbiologic hazards of the aquatic environment. Clin. Dermatol. 5:52.

2. Rao, V.C., K.M. Seidel, S.M. Goyal, T.G. Metcalf, J.L. Melnick. Isolation of enteroviruses from water, suspended solid, and sediments from Galveston Bay: survival of poliovirus and rotavirus absorbed to sediments. Appl. Environ. Microbiol. 48:404.

3. Cavari, B.Z., R.R. Colwell. 1981. Effect of temperature on growth and activity of *Aeromonas* spp. and mixed bacterial populations in the Anacostia River. Appl. Environ. Microbiol. 41:1052.

4. Makintubee, S., J. Mallonee, G.R. Istre. 1987. Shigellosis outbreak associated with swimming. Am. J. Public Health 77:166.

5. Dewailly, E., C. Porier, F.M. Meyer. 1986. Health hazards associated with windsurfing on polluted water. Am. J. Public Health 76:690.

6. Seyfried, P.L., R.S. Tobin, N.E. Brown, P.F. Ness. 1985. A prospective study of swimming related illness II. Morbidity and microbiological quality of water. Am. J. Public Health 75:1071.

7. Coolbaugh, J.C., O.P. Daily, S.W. Joseph, R.R. Colwell. 1981. Bacterial contamination of divers during training exercises in coastal waters. Marine Technol. Sci. 15:15.

Prevention of Skin Problems in Saturation Diving

WILLIAM SCHANE
Box C7, Judith's Fancy
Christiansted, St. Croix, USVI

Introduction

Since 1978, the National Undersea Research Center on St. Croix has conducted 100 scientific saturation missions of 7- to 14-days duration in which 515 aquanaut scientists accumulated about 66,500 saturation man-hours. On average, guest scientists spend about 7 hours of each day in the water conducting their research. In previous in-water saturation operations, prolonged exposure to open sea water has commonly caused serious dermatologic problems, mostly infections, and some have been severe enough to require abortion of the mission but in our own 11 years of experience, we have seen remarkably few skin infections. We attribute this to an aggressive program of preventive medicine. Although we do not dive in grossly contaminated water, it may be possible to apply some of the measures found to be successful in preventing skin infections after prolonged immersions in open sea water to shorter exposures in water contaminated with microbes. Most of the measures are not new or revolutionary.

Procedure

We use a combined engineering/operations/medical approach, and emphasize the importance of proper skin care to aquanauts during their training. We try to encourage compliance by making proper skin care as simple, convenient, and noninterfering as possible.

The skin is remarkably well designed to perform its assigned tasks: to keep those things that should be outside the body out, and to keep those things that should be inside the body in. To perform these tasks properly, the skin should be kept clean, dry, and free from breaks. Our habitat permits maintainence of temperatures within the comfort range for most people; normally it is held between 78° and 82°F, and relative humidity is normally held between 65 and 75%. Clean, dry towels and ample hot, fresh water for showers are available.

The external auditory canal has been the bane of diving medical officers and has been mentioned as a major medical concern in the after-action reports of most saturation dives. To address this problem, the external auditory canal is cleared of all accumulated cerumen and debris before saturation. Macerated epithelium remaining in the canal serves as a mulch which retains moisture and provides a protein-rich, alkaline culture medium ideal for bacterial growth. Excess cerumen acts as a physical barrier to egress of water from the external auditory canal, and because of cerumen's hydrophobic properties, it forms a meniscus with water, which further impairs proper drainage. We believe that a clean, dry, patent external auditory canal is essential before saturation.

After every excursion, our aquanauts are advised to lavage the canal with fresh water. Even if all salt water in the canal is drained, a small amount may remain coating the canal. As the water evaporates, salt crystals will remain. Salt is hydroscopic, and therefore will imbibe moisture from air and keep the walls of the canal moist. In circumstances where the water is microbially or chemically contaminated, lavage with clean, fresh water is particularly important. It is important to dry the external ear parts as thoroughly as possible with a clean, dry, towel and the interior of the external auditory canal should be further dried with a flow of heated, dry air; a hair drier works very nicely.

At the end of every "diving day," the canals are lavaged with specially-prepared ear drops, a mixture of alcohol and acetic acid. The alcohol is miscible with any remaining water, which is then drained away with the alcohol; and the acetic acid leaves behind a pH of about 5.5, the normal pH of healthy skin. The aquanauts are carefully instructed in the proper technique of instillation, and since the procedure takes only a few seconds, we have achieved good compliance.

Using these procedures, we rarely have to delay an excursion for ear problems, and have never had ear problems interfere with scientific productivity. My ear, nose and throat colleagues warn of the potential hazards of alcohol depleting the cerumen and its protective properties but in our hands, the alcohol-acetic acid solution has almost always prevented acute otitis externa.

Intertriginous areas are affected by infection by bacteria, yeasts, and fungi, and by contact dermatitises due to primary irritants. At the end of every excursion, the aquanauts are asked to bathe well with soap and fresh water, dry their skin thoroughly with a clean, dry towel, and don clean, dry, warm clothing. We normally use no other protective measures to prevent infection, and, of course, powders and talcs are not permitted in the habitat.

When a diver spends 3 to 6 hours in the water, it is nearly impossible to avoid urinating into the wet suit. If urine and its breakdown products are permitted to remain in the wet suit, a typical "diaper rash" is common. We recommend frequent and thorough washing of the wet suit bottoms, at least once every other day, using a good detergent, followed by a thorough rinse with fresh water. In addition to reducing the likelihood of a primary irritant dermatitis, this washing reduces the odor of urine in the habitat.

Chafe occurs most commonly on the dorsal surfaces of the toes, behind the knees, in the axillae, along the seams of wet suits, and at the wrists and around the collar line of the wet suit. To prevent chafe, we recommend wearing a full body leotard. We have seen three benefits: the leotard provides an additional sliding surface, greatly reducing chafe; it makes donning the wet suit much easier, particularly when the suit is wet; it increases warmth by reducing water movement under the wet suit. We prophylactically put tape on skin areas which have a high probability for chafe, such as the dorsal surface of toes. It is not unusual for our aquanauts to swim 300 m to and from a work site, so the skin of the toes takes a terrible beating.

We occasionally see traumatic incisions and lacerations of the skin due to interactions between the scientist and residents of the reef or to careless use of tools or knives. To manage these injuries so that the scientist can continue his excursion schedule, we clean, debride, and suture the wound as soon as possible after the injury occurs, usually within an hour, and we try for perfect approximation, especially of the stratum germinativum. The wound is left covered, but not occluded, during habitat time. During excursions, the wound is occluded with a petrolatum-base antibiotic ointment. At the end of the excursion, the wound is washed with soap and fresh water and the ointment removed. In most instances, no prophlyactic systemic antibiotics are used.

Conclusion

In conclusion, skin problems can be controlled in saturation diving by cooperation between habitat engineers, operations personnel, the scientist aquanauts, and the medical staff.

Hypothermia

PIERRE D'HEMECOURT
Hyperbaric Oxygen Center, Suburban Hospital
Bethesda, Maryland

Introduction

Hypothermia has been the scourge of the military since well before 400 B.C., when the Greek general Xenophon attempted to bring 2,000 of his troops through the Armenian mountains and was stopped by snow and freezing cold. In another famous case Napoleon probably lost half his troops during his retreat from Moscow. More recently it was reported that ten percent of the casualties in the Korean War were secondary injuries related to the cold. So hypothermia has been a significant danger for the military. In the civilian world hypothermia really came to light in Great Britain, where many residents of London and other towns were dying of hypothermia, especially the aged and those suffering from diabetes, thyroid, and other health problems.

In the diving community hypothermia also becomes an important consideration. In many dive operations hypothermia ranks in importance alongside decompression because most diving occurs in temperate waters, waters far below body temperature.

Hypothermia results from a decrease in body core temperature below 35°C. There are several ways in which one can lose heat and thus become hypothermic. Cold-water immersion is the most obvious example, but even evaporative cooling can cause hypothermia in disabled persons with decreased thermoregulatory response. There are four mechanisms of actual heat loss from the body. Conduction involves contact between two surfaces, with heat moving from one surface to the other directly, such as would happen in diving injuries where heat is conducted directly out of the body. Convection and evaporation may also play a role in such injuries, but to a much lesser degree. Convection is the transfer of heat via the motion of material, such as wind removing a warm envelope of air, and evaporation is the loss of warm, humidified air. Radiation is the transfer of heat through thermal waves, as when a man ventures out in 4°C air without a hat. A person can lose up to one half his metabolic output through his head.

A body can conserve or gain heat through hypothalamic mechanisms. The hypothalamus receives sensory input from the periphery, as well as receiving the warm blood being directed at it, and can appropriately respond. One of the responses involves immediate vasoconstriction of the vasculature directly below the skin, and this creates an outer shell made of subcutaneous fat and vasoconstricted skin around a warm inner core containing the brain, heart, and the essential viscera. As the temperature begins to drop and the shell becomes larger and larger, the inner core becomes smaller and smaller.

Another heat-maintenance strategy the body uses is to initiate shivering, which causes a five to sixfold increase in metabolic rate. This depletes glycogen storage; thus a person suffering from a glycogen storage disorder may be unable to mount a shivering response. Alcohol consumption also depletes glycogen storage. Below 35°C, the shivering response begins to slow; at body core temperatures below 30°C it stops altogether.

A basic clinical sequence occurs when the body is exposed to cold. As the core temperature begins to drop from 37° to 34°C, there is an initial excitatory phase which occurs with shivering. The A-dynamic phase, between 34° and 30°C core temperature, is a slowing-down phase. Below 30°C begins a paralytic stage where shivering actually goes into a rigor-like state.

The Systems of the Body

Cardiovascular System

The cardiovascular system reacts to hypothermia by increasing output, but as hypothermia continues, blood pressure and pulse drop, as does cardiac output. As for arrhythmias, once one reaches to 32°C, some superventricular arrhythmias may begin, such as sinus arrhythmias and artrial fibrillation, and as the temperature lowers towards the 25°C mark, one can actually experience ventricular fibrillation.

One of the classic signs of hypothermia is the J-wave or the Osborne Wave. Best seen in the anterior leaves, the wave exhibits a little upslope on the QRS complex. At first this was believed to be a precursor of developing ventricular fibrillation. However, it is now understood that it is no more predictive of initiating ventricular fibrillation than is widening of the QRS complex or any of the other superventricular arrhythmias.

Generally, at about 32°C a patient will go into atrial fibrillation. Then he will go into a slow ventricular response with extreme brachycardia. Ventricular fibrillation will occur followed by asystole and death.

Pulmonary System

An initial increase in the respiratory rate causes significant respiratory alkalosis and can actually predispose the person to going into tetany and extreme rigor and decreased ability to swim. Later this effect declines, and the respiratory rate goes down, which is followed by bronchorrhea, decreased gag reflex, and possible aspiration. There are arterial blood gas alterations in hypothermic patients; the pH might appear much lower than it actually is, and the PO_2 and PCO_2 will appear higher. Most importantly, the pH changes must be noted during resuscitation. One will require less bicarbonate replacement than normothermic resuscitation.

Kidneys and Renal System

When the core temperature drops, there occurs a decreased ability of the renal tubules to go through oxidative metabolism, which results in a loss of the ability to concentrate urine. The body loses sodium and water and becomes extremely hypovolemic and hyponatremic. This leads to hemoconcentration, which can become a real problem with thrombosis. For example, myoglobinuria will occur from rhabdomyolosis from the extremities. Hemolysis may follow. This can cause acute tubular necrosis with renal shutdown. This is more of a problem in divers who have been submerged for several hours. These patients are going to be extremely hypovolemic, and during resuscitation and rewarming, this must be taken into account. The victim of prolonged exposure must have aggressive fluid restoration to prevent these problems.

Endocrine System

One may have elevated or depressed glucose levels. The patient may experience decreased insulin output, and glucose levels may increase. On the other hand, increasing utilization of glucose due to shivering may cause hypoglycemia. If the patient is hyperglycemic, this will often correct itself with rewarming, although hypoglycemia may require treatment.

Conditions of Hypothermia

From the perspective of the diver there are three basic situations: mild hypothermia, profound hypothermia, or exposure to severe cold, such as water that is less than 10°C. In a mild hypothermic situation, the patient usually experiences pulmonary excitation, with tremendous respiratory alkalosis which may cause vasoconstriction of the cerebral arteries, resulting in confusion. This can be a significant problem over a short period of time. For the profoundly hypothermic patient, the problems are hemoconcentration, hypovolemia, and dehydration. The utilization of glucose due to shivering may cause hypoglycemia. If the patient is hyperglycemic, this will often correct itself with rewarming, though hypoglycemia may require treatment.

When patients have been exposed to extremely cold temperatures such as sudden immersion in frigid water, there is initially a tremendous output of catecholamines, and these people go into cardiac arrest instantaneously. If they survive that instant shock of going into the water, a loss of gag reflex may cause aspiration. Interestingly, a paradoxical vasodilatation of the subcutaneous vasculature may occur rather than the vasoconstriction.

There are certain predisposing factors for hypothermia, such as age. The very young, because of a higher surface area-to-mass ratio, can become hypothermic very quickly. They also have less subcutaneous fat, especially young boys, who have more of a chance of becoming hypothermic than young girls. The very old, for almost the same reason, "less subcutaneous fat, less muscle," have more problems with hypothermia. Decreased endocrine function may not allow them to mount an effective response. Barbiturates or alcohol also increase risk, since these drugs cause decreased vasoconstriction in the periphery.

Prevention is most important. Adequate insulation and good physical fitness are primary. People who swim the longest in water, such as the long-distance channel swimmers, have a relatively large amount of subcutaneous adipose tissue, yet they are in very good physical shape. People who have a lot of adipose tissue but are not in good physical shape cannot mount the shivering response and the other acclimatization responses to withstand long-term swimming in cold water. Acclimatization allows the neurohormonal system to cause vasoconstriction more easily and causes more extensive areas of shivering.

If there is an accidental immersion in water, thermal float jackets are useful in keeping the head above water and decreasing heat loss from the head. The fetal position is considered one of the better positions because the knees are drawn up and the arms are together, decreasing heat loss from the groin and axillary (armpit) areas. One should avoid swimming because this will increase heat loss.

Treatment

Passive, active external, and active internal rewarming processes can be applied. The passive course is simply to remove the patient from the frigid environment, wrapping him or her in blankets and any other kind of material that will allow internal regulatory responses to work. Active external heating consists of placing heated objects on the body because the body can't mount the internal regulatory response. Active internal heating is the use of more invasive and some not so invasive techniques such as warm-water nasogastric lavage.

The problems with active external warming are threefold: if a patient has been in the water for a number of hours and is severely hypovolemic, active external warming could cause hypotension very quickly. Furthermore, a significant afterdrop effect will occur with vasodilatation of the periphery bringing cold blood into the central core area. A third problem results when the cold blood returns loaded with electrolyte abnormalities, such as lactic acid. Of these three modalities, the one that causes the most concern is the active external. Both active internal and passive warming are the safest methods used, both in field and hospital situations.

As a rule, the field treatment for hypothermia is passive, if the patient is able to mount his or her own thermoregulatory response, has alert breathing, and has his/her own pulse. Active external methods are used in situations of short exposure where there is minimal hypovolemia. Active internal methods are usually not available in the field, except for breathing warm oxygen or air at about 42-45°C. There is some controversy in the literature as to whether oxygen or air should be used, with some reports showing that 100 percent oxygen causes increased pulmonary and cerebral problems. The best approach is probably to use 50-percent oxygen mixtures.

Other active internal methods employ an IV solution warmed to about 42-50°C. Two methods of warming saline are placing the bagged solution in a warm tub of water or putting it in a microwave oven. However, never use a dextrose solution in a microwave as it tends to crystallize when exposed to microwave radiation.

Peritoneal lavage is an extremely effective way of rewarming. Roy Myers of Baltimore did a study in which the two methods he found most effective were lavage and peripheral circulation, for which one could use some type of bypass circuit. He was able to show significant increased warming under those circumstances.

Regarding airway management, there has been much written about caution in handling the airway because one may stimulate ventricular fibrillation. Indeed the cardiac status is very irritable in hypothermia. However, the same principles for intubating (using tubes with) a normothermic person should be followed. If the patient is not breathing, intubation is necessary. If the patient is somewhat slow in his or her breathing and has a pulse, careful observation is advised. However, the advantages of intubation are the use of warmed O_2 for central core rewarming and protection of the airway from aspiration. If CPR is being used one may hold back on external chest compressors and ventilation about 50 percent to avoid stirring up the myocardium. One generally does not need that high metabolic rate at that time anyway.

As for physiological monitoring, one has to watch the vital signs. Labwork must be monitored carefully. It is better to draw your blood from arterial rather than venous sites, because the veins are returning blood from the periphery in a markedly abnormal state. An extremely hypoglycemic patient requires immediate treatment. Acidosis does not require treatment unless the pH goes below 7.0.

The medications for the ACLS protocol such as dopamine, lidocaine, etc., are not very effective during severe hypothermic conditions. For significant ventricular instability, bretylium is the most appropriate drug to use. Defibrillation is probably not effective below 28°C.

Conclusion

From a resuscitation standpoint, my rule of thumb is that no one is dead until he or she is warm and dead. This basically means that the rescuer has to bring the patient's temperature up to around 30° or even 33°C before they can conclude that the individual is no longer resuscitatible.

Unsolved Problems of Undersea Living

LAWRENCE W. RAYMOND
Medical Director
Exxon Research and Engineering
Florham Park, New Jersey

Introduction

It is a little pretentious for a diver not in the trenches of diving physiology and medicine to discuss unsolved problems Some might say the biggest unsolved problem is lack of adequate resources, not just of money, but of people. If there were a service that I could do for diving physiology and medicine, it would be to spend more time with freshman medical students.

Hyperbaric Problems

There is a lot to learn about diving physiology, about CNS, about pulmonary oxygen toxicity and red cell oxygen toxicity. We have not seen much information about renal oxygen toxicity since an early report in *Pennsylvania Academy of Sciences Proceedings*, but it is probably there and will resurrect itself in due time. I also continue to be reminded of the problems with nitrogen narcosis. A diving partner of mine once did a 285-foot air dive, which used to be in the Navy training tables and I hope still is. When, upon reaching the bottom, he was asked for his social security number, he replied, "I am number One!"

Around the time that I was born, it was discovered that helium-oxygen diving solved a lot of problems at intermediate depths, 150 feet and deeper, which I believe is still the point at which commercial divers will not air dive unless they have to. In such a case, it is certainly very comforting to know that one can dive on air and function under the control of the Master Diver to, let's say, around 250 feet.

One practical consideration is basic physical fitness. This remains essential. If anything that can go wrong at one atmosphere goes wrong underwater, the diver has that much more trouble.

For many, diving is an occupation. Hull inspection may be a fairly humdrum businesss, but it has its hazardous moments too. Working divers may be confronted with challenges that are very different from those seen by the person who does all his or her work around hyperbaric chambers. I won't review the rationale for saturation diving. It has been a quarter century since Captain Cousteau, speaking at a symposium at Duke University, gave credit to Captain George Bond of the Navy for pioneering this approach to streamlining decompression from very deep diving.

Thermoregulation

I would like to focus on some different perspectives on thermoregulation, most of which will tie in with observations made by Dr. Hong, a leader in this scientific area. Respiratory insufficiency and weight loss and diuresis really should not be a question because diuresis is a big part of the hyperbaric phenomenon.

Divers do come in all sizes and shapes. When I was actively involved with the Navy, there were many divers who looked a little heavy, and were a little heavy, but they were powerful, muscular men. The ones who spent a lot of time in the water did seem to develop a "pannus" of self-insulation. Unfortunately, as Hong has reminded us, what's good in terms of insulation may not be so good in terms of decompression. Those who have this extra layer of endogenous protection probably are at a little bit greater risk of diving decompression accidents than their more slender colleagues, but for cold-water diving, it's a reasonable trade-off.

Consider a diver exposed to temperatures of 25°-28°C (in the Chesapeake Bay in the summer it doesn't get much higher than that if one gets away from shore): in other words, a very mild challenge to thermal homeostasis. A rather lean diver starts out with an adequate rectal temperature, but in the course of the repetitive cycle of rest, work, recovery, surface, in about an hour's time he has lost 2°C even though the water temperature is very mild by diving standards. The dive is to only 10 feet, so the diver can come up, have blood drawn, then go back down for another cycle, and he is wearing a wet-suit top which is taken off to draw the blood. We observe a progressive rise in catecholamines from very low levels in the predive resting state to substantially elevated levels.

Clayton at Tripler, in cooperation with Hong and others, has studied the impact of head-out immersion on various aspects of cardiopulmonary function. One can look at the impact of immersion at various hypothermic temperatures. The work at Buffalo looked at 43°C temperature, 35°C euthermal, and 20°C, which is a modest hypothermic challenge. We know that this will centralize blood volume; immersion does that. Yet in addition, superimposing the peripheral nasal constriction of mild, moderate, or severe hypothermia accompanies cutaneous vasoconstriction. That further centralizes blood volume, and what does that do to gas exchange in terms of pulmonary capillary volume and transit times through the capillary net?

I believe this represents an area for further investigation. I can find nothing in the literature about this kind of experimentation, which uses the total respiratory resistance unit forced-oscillation methodology. Yet it would be a fairly simple experiment that would lend further insight. Does the lung undergo an erectile type of response because the submajor blood vessels are enveloping the airways? Does this cause changes in airway conductance that could be demonstrated with this technique? These are interesting questions for which such experimentation might well provide the answers.

In the 1970s we believed that an electric suit would prevent hypothermia. At that time we thought electricity would solve the problem, but perhaps we were too optimistic. Heat is lost from the electric suit via breathing tubes without heating either the inspired gas or providing some different kind of breathing gas insulation. The impact of breathing very cold gases has had some attention, and further investigation would be appropriate unless we can protect the diver from inhaling gases close to ambient temperature in the ocean. I believe that there is a Navy requirement that on dives deeper than 600 feet, breathing gas be heated anyway, but again I would look for a lot of the physiological research in this area. I also pose this question: if one is providing gas, hot helium with a high heat capacity, what is that going to do to such things as mucociliary clearance? Sometimes a quick fix to a problem may not represent a total solution.

Solving the thermal comfort problem via high ambient temperatures presents at least one unstudied problem, that of raising the temperature of the scrotal contents. This rise in scrotal temperature may be enough to interfere with spermatogenesis, a thermally sensitive function. If we are serious about prolonged human habitation underwater, we will have to take a look at this possibility. The quick fix for raising the ambient temperature probably will not do it. I am not sure if this is specific for helium. We all talk about how helium has terrific conduction properties. John Severinghaus once reminded me that heat transfer is not just a convection coefficient. It is related to density times specific heat, whether one is dealing with helium or any other gas. If we had people living for prolonged periods in the deep we might have to consider scrotal temperature. If one turns off spermatogenesis, one would want to assess its return in careful detail. Again, these are some problems that we could encounter in the future.

Respiratory Problems

Bruce Templin has attempted to introduce respiratory functions in deep diving, executing a maximum expiratory flow maneuver with a wedged spirometer. The control measure was done in air at one atmosphere topside, at the Taylor Diving Salvage in Belle Chase, Louisiana. A report that was done with Ed Lanphier and company included the present study results, which were very close to those of Bruce Broussolle. If we dissect the flow-volume loop, we find that it is a little crowded. We see that the flow rates are diminished at all lung volumes as one goes deeper. There is some hysteresis here. During compression one tracing is obtained, and during decompression, another. Again, that may well be a training effect. One of the better papers on training that I recall is one by Mark Bradley and Gerry Meade showing that by endurance maneuvers, one can improve people's ability to do MVV maneuvers. Probably a similar effect on training and respiratory muscles is at work in these divers.

Now, for those interested in something that gets a little closer to what divers do when they have to work, suppose one has a diver who becomes unconscious. Trying to get the diver up to the surface for resuscitation, is going to be very hard work indeed! This is just one example of an exhaustion problem where exercise ventilation might well approach maximum voluntary ventilation.

Lanphier's early work predicted what would happen to MVV at depth. Again we see a hysteresis, and other workers followed the same curve in this relationship. Once divers get down this deep, they don't do a whole lot worse by increasing density to 17 grams. Most of the penalty is paid in the first thousand feet or so of the descent. The denser gas brings out a tendency in some individuals to retain CO_2. This tendency is increased by exercising, even with minimal breathing gear, Lanphier showed.

Recent findings in divers and former divers suggest that it is hard to predict which subjects during exercise will retain CO_2. This is another unsolved problem. A group in Buffalo has data on subjects that run PCO_2 up to 78 instead of the normal 40 or 36 PCO_2.

Divers who registered well on the ergometer, when working out at sea level, ran into difficulty when they attempted the same exercises at 49.5 atmospheres. They found themselves ready to pass out, saw red, took in big deep breaths and discovered it did little for them. In short, they were unable to perform anywhere near as well as what they did at sea level. Lambertsen made similar observations with younger men who did much more exercise at depths. He said, "I don't know what was wrong with those EDU divers, but we didn't see it."

Experimentation

A lot of the research on deep diving, of course, is done in hyperbaric chambers. It has to be that way. Many experiments could not be performed on the open sea where lack of control over the surroundings could produce catastrophie in the event of anything going wrong. Occasionally a mishap in the open sea might cause one to retreat back to the hyperbaric chambers and restudy them.

Bill Spaur's design was an "ark within an ark", a plastic swimming pool built within the confines of a hyperbaric chamber and equipped with an ergometer. A small porthole penetrated the wall of the ark so that the diver could have his radial artery cannulated. A diver on the ergometer wore a Mark diving hat, and outside of the plastic enclosure, but still inside the hyperbaric chamber, a tender drew the arterial blood samples. The diver had to stick his arms out of the boot and have the tender in the dry environment draw the blood sample. After the analysis topside, we would have a very prompt indicator of the diver's condition. Blood samples were very carefully tonometered with the university-developed techniques in the hope that when we set our

PCO_2's for 38, we would know that they were plus or minus 0.5 mm Hg.

With these hyperbaric chamber exposures, one would expect that CO_2's might be elevated, and I was sure they would be. But we found the opposite. Yet breathlessness was intense: it was work-limiting and possibly life threatening in the open sea, but it was not associated with high CO_2's in our divers. In fact, even their lactic acid levels were at very modest elevations during very modest work rates compared to what they could do at sea level. I think this represents yet another unsolved problem.

Weight Loss

As for the problem of hyperbaric weight loss, in our experience, this represented about a 5% drop in predive body weight, which was not explained by caloric imbalance. It was not explained totally by the diuresis effect either, though that explains a lot of it.

Striking hemoconcentration goes along with the hyperbaric diuresis. Serum albumin, which rises along with protein, was observed to rise over 1 g/100 ml. Other researchers have worked in the interim on some fine studies that have elucidated most of the mechanisms of hyperbaric diuresis. However, I think there still remain some unanswered questions because, in their hands, antidiuretic hormone, or vasopressin if you will, dropped.

Work done with Carolyn Leach of NASA shows a modest increase in antidiuretic hormone (ADH) and a modest but significant blood hemoconcentration at the time of greatest weight loss in our divers; i.e., at the time when their serum sodium has risen and then dropped. Total protein rises strikingly, as does antidiuretic hormone, as it might logically be expected to, in defense of intravascular volume. Of course, as we are all aware, if the body loses enough volume, it releases ADH further to defend blood volume to avoid becoming more hypovolemic. I will not go into cardiovascular consquences. That again needs more work than it has received. Dr. Lin has shed some important light on that, but from the perspective of some of the young men and women who are going to be divers, I would like to know more about the subject.

These findings may seem discordant. Everything everybody knows about antidiuretic hormone in hyperbaria suggests that hormone levels go down. The fact is that antidiuretic hormone concentration goes down until blood volume becomes so depleted that it stimulates an increase in the hormone. In the face of hypovolemia and increased ADH, divers continue to elaborate a dilute urine in comparison to the predives and the postdives. And this represents yet one more of the many unsolved questions pertaining to deep diving.